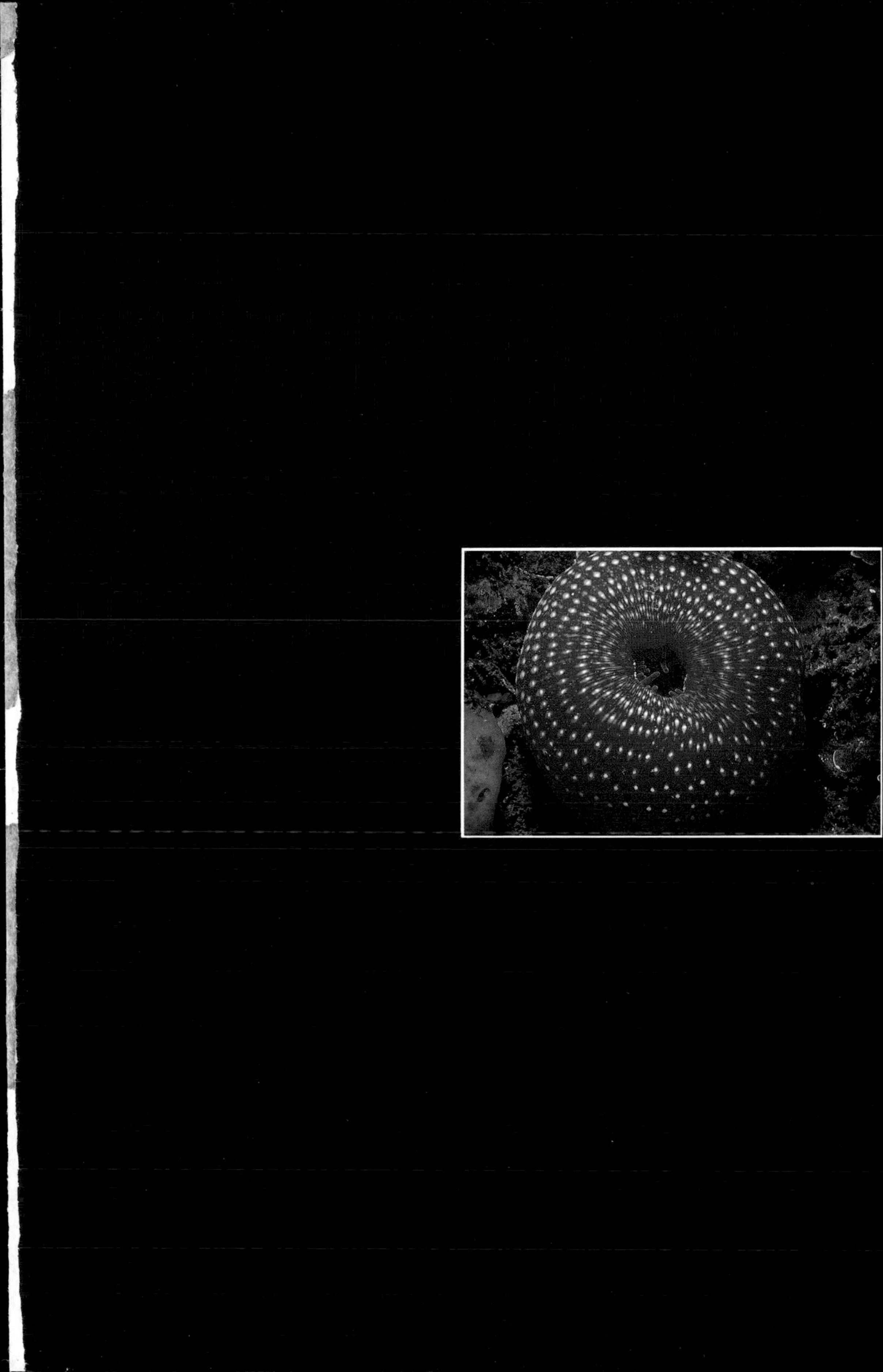

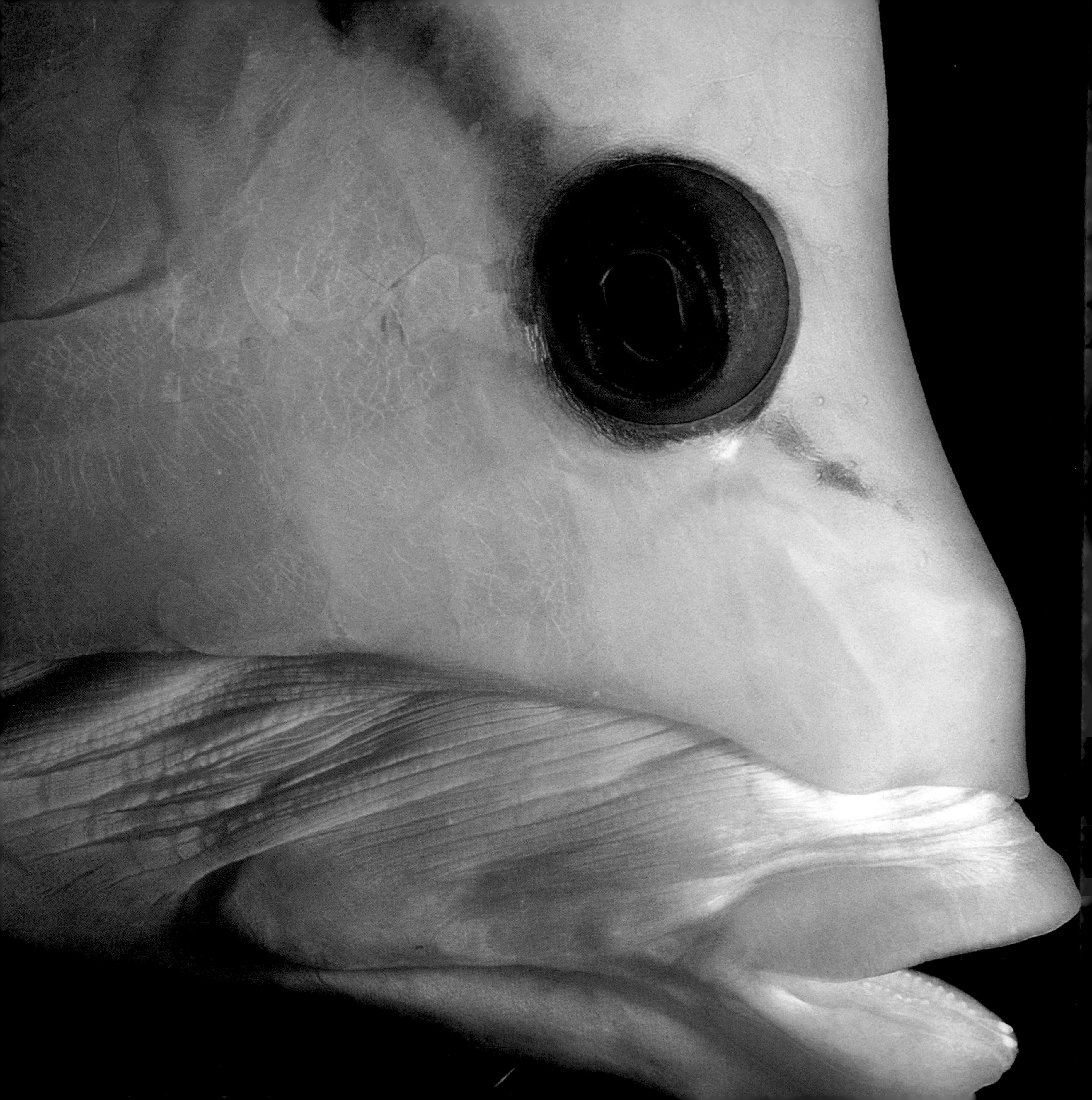

Undersea Life

Text by Joseph S. Levine
Photographs by Jeffrey L. Rotman

Stewart, Tabori & Chang, Publishers
New York

To our parents
Renee and Nate, Carol and Bob

Published by Stewart, Tabori & Chang, Inc.,
740 Broadway, New York, New York 10003

Sections of chapter 4 have appeared in a different
form in *Science Digest, Animal Kingdom,* and *Geo.*
Sections of chapters 5 and 6 have appeared in a
different form in *Smithsonian.*

Library of Congress Cataloging in Publication Data

Levine, Joseph S., 1951–
Undersea life.

Bibliography: p.
Includes index.
1. Marine ecology. I. Rotman, Jeffrey L.
II. Title.
QH541.5.S3L47 1985 574.5′2636 85-2833
ISBN 0-941434-70-2

85 86 87 88 10 9 8 7 6 5 4 3 2 1

Distributed by Workman Publishing
1 West 39 Street, New York, New York 10018

Printed in Switzerland

First Edition

Page 1:
California sea anemone deflated into a compact mound.

Pages 2–3:
California sea anemone opened wide to feed.

Pages 4–5:
Glassy sweepers.

Page 6:
Epibolus, the sling-jawed wrasse.

Contents

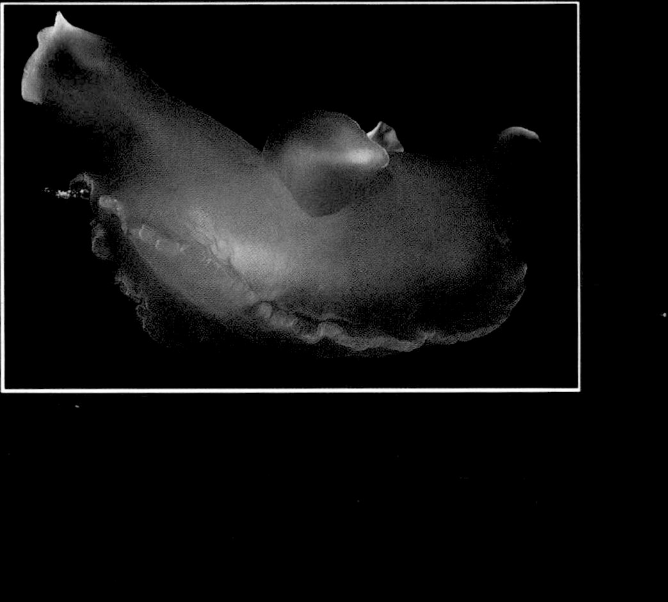

Introduction

Humanity has entered a period of unprecedented change in its relationship with the oceans. Knowledge of the sea and its workings is expanding faster than ever before in history. Deep-sea submersibles visit the deepest ocean trenches, and shallow-water undersea habitats house investigators for weeks. Recombinant DNA techniques probe the innermost intricacies of single living cells, and computer-based manipulation of massive data bases allows analysis of global phenomena once well beyond our grasp.

This knowledge revolution is philosophical as well as factual, for marine scientists are systematically abolishing the traditional, artificial barriers between academic disciplines and are concentrating instead on the critical interconnections among these disciplines. As ecologists, physiologists, behaviorists, and fisheries biologists have entered dialogues with each other and with oceanographers and meteorologists, they have uncovered layer upon layer of interrelationships among marine animals and plants—and among ecosystems once viewed as independent entities.

So interconnected are the world's ecosystems that, to some students of global ecology, the workings of all living organisms on earth together resemble the life processes of a global superorganism. Within this great, hypothetical being—called Gaia after the Greek name for the earth—energy flows and nutrients cycle through atmospheric and oceanic circulation patterns that serve as the living planet's bloodstream. Like organs on whose function the existence of ordinary animals depends, certain areas of the earth fulfill essential roles in food production, others in reproduction, and still others in recycling wastes. Each ecosystem, each microhabitat, each population, and each species on the planet plays a critical role in Gaia's life processes.

This is the world view that Jeff Rotman and I have advanced in *Undersea Life.* The textual and photographic images of animals and ecosystems are presented not simply as a collection of odd and intriguing creatures and habitats but as integrated components of a living net whose finest filaments include the genetic code and whose expanse encompasses the globe.

Hidden in coral crevices by day, the Spanish Dancer nudibranch emerges to feed and swim only after sunset.

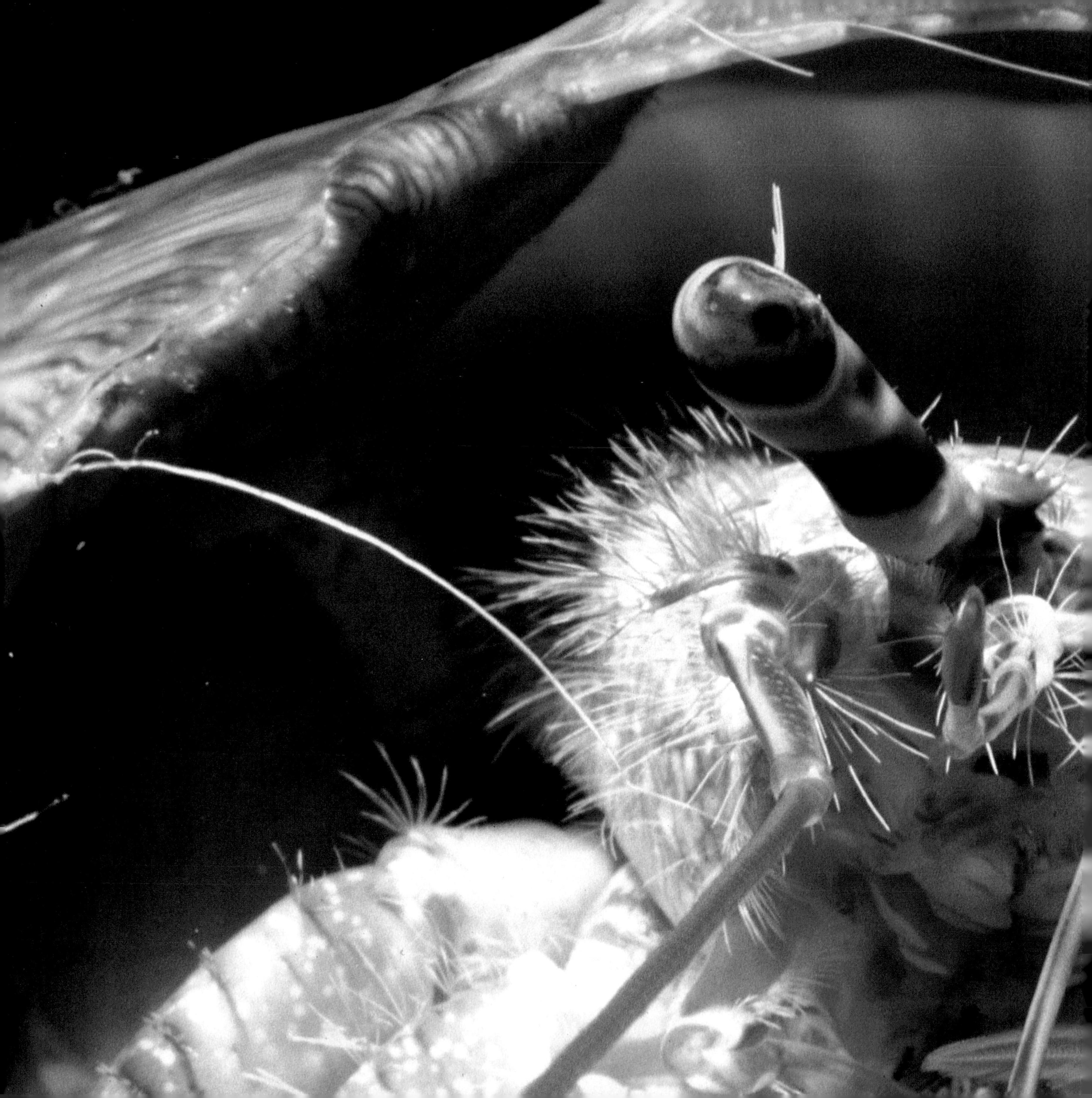

The structure of the net is determined by the basic principles of life itself: the capture of energy, the cycling of nutrients, and the unpredictable and creative process of evolution that has shaped the living world around us. We open the book with introductions to these concepts because it is only by understanding the universal principles that unite all living things, and by recognizing the inherent differences between terrestrial and marine environments, that one can fully appreciate many of the fascinating details of undersea life. And only with an intuitive feel for the workings of evolution can one comprehend the intricacy of the ecological balance in marine systems.

From these basics we turn to examinations of several important marine ecosystems. These ecosystems were chosen not simply because they are beautiful and fascinating but to demonstrate how various organisms satisfy life's basic requirements in habitats ranging from the cold, turbid North Atlantic to the tropical Red Sea. Time and again, we show how the apparently idiosyncratic techniques utilized by marine animals for feeding, reproduction, camouflage, offense, and defense make surprising sense when examined as evolutionary responses to the physical environments in which those organisms live and the biological environments which they create for each other.

It soon becomes apparent that, just as cells are parts of organisms, organisms are parts of ecosystems, and ecosystems parts of the biosphere. To illustrate these interrelationships, we point to the critical role played by spawning and nursery grounds like George's Banks in the health of the fisheries of the Gulf of Maine. We show how salt marshes and mangrove swamps are bound to wide-ranging offshore fish populations and stationary coral reefs alike, both by physical processes and by the regular migrations of feeding and spawning animals.

Throughout the oceans' living net is woven the decidedly terrestrial thread of human interrelations with the sea, and *Homo sapiens* finally emerges to dominate the last chapter. Humans interact with marine organisms and ecosystems in ways that are often unexpected and always intriguing, especially when viewed from a perspective informed by a grasp of marine ecology. Exploring the context in which marine organisms have evolved poisons and venoms for their own purposes will give you a better idea how some of these compounds can serve as both weapons in the struggle against disease and cancer, and as tools for medical research. Understanding the way naturally occurring chemicals pass from one organism to the next in marine food chains will help you to see all too clearly how manmade chemicals introduced into the environment can ultimately affect our bodily processes and those of other animals around us. From disastrous cases of overfishing to the promises of marine aquaculture, you will glimpse both major successes and major blunders in mankind's ongoing interactions with undersea life, and you will comprehend—to the extent current knowledge allows—how and why things have gone wrong.

But beyond the immediate concerns and short-term pressures of extracting food and useable goods from the sea, our book also addresses the critical need for sensible management of

Preceding pages:
Jointed mouthparts at the ready, and antennae aquiver, this hermit crab searches for food on the nighttime coral reef.

the world around us, echoing the clarion call of biologists everywhere for respectful attention to the needs of our planetary organism.

Buckminster Fuller spoke often of "spaceship earth" to emphasize the globe's finite nature and to reinforce the need to tend our environment as carefully as we control the conditions around our astronauts. But the biosphere as it exists today is in no way comparable to any artificial device; no mechanical contrivance known to man could function while its component parts appear, disappear, and change their functions randomly. Yet, although earth's ecosystems shift and change, and although species evolve and become extinct, the biosphere, as a living system, is able to grow, repair itself, and maintain itself in equilibrium. No, mother earth is not a spaceship, regardless of how earnestly our technocracy might wish to see her as one. She has no easily accessible master control panel, and the living programs that direct her life processes are as yet beyond our power to alter.

The position of *Homo sapiens* as a component of Gaia can be seen—depending on our species' future actions and one's point of view—either as the biosphere's emerging brain, or as a vicious cancer. By choosing the former role, acting as a self-regulating center and exercising intelligent control over its interactions with the biosphere, humankind can perpetuate a vital, living planet. By choosing to act as a cancer, continuing to grow, feed, and pollute without restraint, society may ultimately undermine Gaia's metabolism and leave its own descendants with a planet that will need to be managed as tightly and as artificially as a spaceship—a job human science may or may not be able to handle.

Humanity's definitive role in Gaia's evolution will not be decided in any of our lifetimes, but each of us must shoulder the responsibility for the path human society ultimately takes. Educators and scientists always labor under the assumption that increased knowledge leads to increased awareness, which in turn, leads to sensible action. If we have done our job properly, the combination of visual imagery and information this book contains will delight and amaze you. Some of it should provoke you into thinking seriously about the wisdom of mankind's current habits in dealing with the sea. Much of it should raise questions that have no simple answers. Hoping that our book will encourage you to search for those answers, both in the scientific work of others and within your own conscience, we invite you to explore with us the mystery, beauty, and intricacy of undersea life.

To Live in the Sea

Capturing the Sun's Fire

Goatfish hunt with tastebud-laden barbels; wrasses await any fleeing sand dwellers.

In the cold, clear waters off California's coast, a 9-meter-high forest of yellow-green kelp sways gently in the current. Ten meters beneath the surface, light from a photographic strobe reveals a crust of pink coralline algae clinging to rocks, shells, and any other exposed, hard surface.

In warm, shallow bays along the Grand Bahama Banks, emerald-green turtle grass carpets the sandy bottom.

Beneath the low tide line on the rocky shores of New England, a species of oxblood-red algae called Irish moss covers the jagged rocks.

Half a world away, off the coast of Peru, floating, single-celled algae grow and divide in a daily rhythm as the currents sweep them along.

Kelp, grasses, encrusting algae, and floating algal plankton are the oceans' primary producers. Among all marine life, only these photosynthesizers can harness solar energy to assemble inorganic molecules into the complex building blocks of living tissues. Virtually all animals depend on primary producers for the energy that makes growth, movement, and reproduction possible. Animals can store energy in the form of fats, oils, or proteins, but they can obtain this energy only by eating plants that have manufactured sugars, starches, and carbohydrates—or by eating other animals that have eaten those plants. Animal life, constrained by this dependence on plant compounds, must structure its existence around the need to obtain them. Plant-eaters must live within reach of growing plants. Animal-eaters can survive only where other animals are available as food. Thus, any conditions that determine which primary producers can grow in an ecosystem indirectly control the lives of the system's herbivores and carnivores, as well.

Reality is actually far more complicated than this simple picture of natural supply and demand. Animals rarely settle for just any source of energy; many are as particular as the most fastidious human gourmet, constantly picking and choosing among available foods. And just as animals must rely on plant compounds, plants are utterly dependent on solar energy and a steady supply of inorganic nutrients. For these reasons, the structure and dynamics of every ecosystem depend ultimately on the amount of solar energy and nutrients available to the photosynthesizers. Knowledge about energy and nutrient distributions is therefore indispensable for understanding the idiosyncrasies of marine ecosystems and their inhabitants.

The importance of energy and nutrient availability is not intuitively obvious to terrestrial creatures because continents and oceans differ in major ways as habitats. Seen from space, the earth's land masses—with the exceptions of their deserts—are colored green by the primary producers that cover so much of their surface. The sun beams at the earth's surface a constant stream of electromagnetic energy, ranging from ultraviolet wavelengths shorter than human eyes can see to infrared wavelengths longer than we can see. Because the atmosphere transmits most of these wavelengths freely, all terrestrial environments, from lowland savannas and valleys to alpine meadows, receive more than enough light for photosynthesis. In most environments, too, soils contain ample nutrients to support plant growth.

Life for plants beneath the ocean's surface is an entirely different matter. Water, even clean water, absorbs and filters light, removing much of its energy and changing its color. Dissolved and suspended materials further affect the color and intensity of light that penetrates to any depth by absorbing and scattering certain wavelengths more than others. Just how much and in what way light is altered depends on both depth and location.

In the open sea and near desert continents or small islands where land runoff is minimal, oceans are relatively free of organic matter. In such places, sea water absorbs red and violet wavelengths strongly enough to effectively eliminate most energy in these parts of the spectrum within 25 meters of the surface. Sea water in these environments transmits blue light better than any other color. This, therefore, is a world seen not through rose-colored glasses but through blue ones, for blue light alone penetrates farther than 75 meters, bathing the underwater scenery with an eerie, turquoise glow.

Along temperate seacoasts, dissolved organic matter from fresh-water runoff combines with high concentrations of phytoplankton to remove a good deal more blue light. Here, light of all wavelengths is absorbed more quickly than in clear water, and the especially strong blue absorption causes yellow-green light to predominate within even a short distance of the surface. Here the illumination is far dimmer, and is usually colored somewhere between pea green and chartreuse.

These theatrically impressive changes in underwater lighting put marine primary producers in a serious bind. For as cool and appealing as blue and green light may be to human eyes, both are nearly useless to most plants. Chlorophyll, the living world's primary solar energy converter, absorbs and harnesses most efficiently light in the violet and red regions of the spectrum—precisely those colors removed by sea water. Although algae and plants have evolved three different kinds of chlorophyll, each of which captures energy from slightly different parts of the spectrum, none of the three by itself can absorb enough light to permit rapid photosynthesis very far beneath the surface.

In response to this challenge, marine algae have augmented their energy-collecting abilities by evolving other compounds—accessory pigments—that assist their chlorophylls in collecting sunlight in the difficult underwater-light environment. Typical land plants and many algae look green to the eye because their chlorophylls absorb red and violet light while reflecting the green wavelengths they cannot use. Accessory pigments absorb some of the otherwise useless middle wavelengths, simultaneously making more energy available for photosynthesis and endowing marine algae with a range of striking colors. Some accessory pigments absorb blue light most strongly. These pigments, usually chemical compounds called carotenoids, combine with chlorophylls to dye their bearers yellow-green. Other accessory pigments absorb both blue and green light; pigments such as these—called phycoerythrins (*phyco*= algae; *erythro*= red)—tint the algae that contain them dark red or purple. Still other algae load their tissues with compounds that absorb light across most of the visible spectrum, coloring the plants a dark greenish brown.

Gorgonian coral polyps, held at right angles to
prevailing currents by fanlike arrays of branches,
snare passing plankton.

But although the light these algae reflect turns them into a living Persian carpet for us and other animals to see, it is the light they absorb that matters to the algae themselves. Accessory pigments, using a bit of electrochemical legerdemain, pass absorbed energy to the algae's photosynthetic machinery. By putting this otherwise wasted energy to work, accessory pigments allow their bearers to photosynthesize and grow at depths substantially below the level at which chlorophyll alone fails to trap enough energy to support growth and reproduction.

Even with these clever devices to assist them, however, marine photosynthesizers fight a losing battle to colonize deepwater habitats. Sea water removes so much energy from sunlight that even the most well-equipped photosynthesizers can only operate relatively close to the ocean's surface. The region within which photosynthesis can occur —the photic zone—may be as shallow as 30 meters in the turbid waters of the North Atlantic or as deep as 200 meters in the crystalline tropical Pacific. Only in this uppermost region, little more than a thin skin on the face of the deep, can floating phytoplankton grow actively, and only where the ocean floor is within this depth along continental margins can plants and the vast majority of algae attach themselves to available surface on the bottom and grow.

Until as recently as the end of 1984, biologists believed that no photosynthetic organisms of any kind could grow permanently at depths below 200 meters because of absolute light limitations. But in January 1985, a research team manning the Johnson Sea Link submersible discovered a community of red algae thriving as deep as 268 meters below the surface near the summit of a previously unexplored Caribbean seamount. It is likely that similar deepwater algal communities exist in other clear tropical seas, but their total extent and their contributions to marine ecosystems are mysteries yet to be unraveled.

These deepwater pioneers aside, the limit to photosynthetically useful energy in the sea causes marine organisms to distribute themselves in patterns dramatically different from those taken by their terrestrial counterparts. Beneath the thin photosynthetic layer lie vast volumes of water through which animals may move, but in which no plants can grow, either in the water or on the bottom. To imagine an analogous situation on land is nearly impossible. If our atmosphere absorbed light the way sea water does, terrestrial plants could exist only at the tops of the world's highest mountains and in a floating layer carried by the winds somewhere up in the stratosphere. Humans, as surface dwellers, could either eke out a living by farming on the plateaus of the Andes and Himalayas or scavenge for scraps of whatever dead plants and animals tumbled down from those regions. The vast plains and green, rolling hills of our continents would be as dark and devoid of plant life as a tomb.

But the capture of the sun's rays by photosynthesizers is just the beginning of the energy story. From algae and plants, this vital energy flows to herbivores, and then to carnivores. It travels still farther, to carnivores that eat other carnivores, to parasites that prey on everyone, and so on.

Each step in this multilayered system of organisms eating other organisms is called a trophic level. There is not necessarily any limit to the number of trophic levels in an energy chain, but the farther away from the primary producers the chain extends, the less energy remains. Every time one organism eats another, an enormous amount of energy is lost. Finding food requires energy. Respiration, growth, and repair require energy. Digesting food requires energy. And very few animals come close to extracting all the energy contained in the food they do digest.

These losses severely limit animals' ability to utilize energy from trophic levels that precede

In shallow, sunlit waters, seagrasses photosynthesize rapidly.

Accessory pigments in purple coralline algae aid chlorophyl in harvesting underwater light. Encrusting colonial hydroids use algal stems for support.

theirs. Although estimates of energy transfer vary widely, on the average only about 10 percent of the energy fixed by plants is available for use by herbivores. Only 10 percent of the energy that herbivores accumulate is available to the carnivores that eat them. And only 10 percent of that energy—10 percent of 10 percent of 10 percent of the original amount—is available to higher-level carnivores.

Energy thus flows through ecosystems, from one trophic level to another, in a steady, unidirectional stream. As it flows, it powers the life processes of all the animals it touches, like a mountain stream driving a mill wheel. But unlike a river, which grows larger as it flows away from its source, the energy stream gets smaller.

The less energy available at any trophic level, the less living tissue that level can have. For this reason, wasteful energy chains usually create what are called ecological pyramids. In an ecological pyramid, each trophic level contains only a tenth of the amount of living tissue the layer preceding it contains. Often, this relationship also produces a pyramid of numbers. For example, a researcher is apt to find roughly ten times as many shrimp in an estuary as there are herring to eat them. Because many marine primary producers are quite small and are eaten by small grazers, many large, commercially valuable animals are separated from their original energy source by complicated food webs containing many trophic levels. Both the length and the complexity of these feeding relationships can severely limit the harvesting of large marine organisms for food and other purposes.

The Staff of Life

As members of each trophic level consume members of the level below them, they acquire the energy essential to life by ingesting organic molecules carrying that energy. But though the transfer of energy and of organic nutrients from one trophic level to the next occurs in a single process, the total picture of energy movement and nutrient movement in ecosystems is quite different. Energy arrives steadily from the sun and flows from one trophic level to the next, powering life processes and disappearing as it goes. But no inexhaustible fountain of nutrients supplies the earth with the atoms that form living molecules. These nutrient elements are neither created nor destroyed; they are passed around from organism to organism in tight circles.

Many nutrients required by living tissues are known to every gardener. Nitrogen, phosphorus, and potassium—needed by all forms of life in relatively large quantities—are contained in every garden fertilizer. So are manganese, iron, and zinc. Plants take carbon from the atmosphere in the form of carbon dioxide. Trace nutrients, such as iodine and sulfur, are needed in much smaller quantities. These nutrients may be toxic in large doses, but their presence in smaller amounts is nonetheless essential to proper growth. These various elements unite in different combinations to form proteins, DNA, RNA, and other molecules such as sugars, carbohydrates, fats, hemoglobin, and chlorophyll.

Nitrogen is a good representative element to follow on its cyclical path through the ecosystem, for most essential nutrients have cycles at least as complicated as nitrogen's. Using energy from photosynthesis, primary producers absorb and concentrate simple, nitrogen-containing compounds such as ammonia from the environment. Additional solar energy fuels the assembly of nitrogen, hydrogen, carbon, and other elements into proteins and amino acids. When animals eat plants or other animals, their digestive tracts break apart proteins in the food and liberate individual nitrogen-containing amino acids. Some of these amino acids are reassembled into the personal proteins of the animal that has eaten them. Others, broken down to release the energy they contain, release nitrogen in the form of ammonia.

Because ammonia is highly toxic to cells, even in low concentrations, it must be eliminated or converted to another, less poisonous chemical form as quickly as possible. Most marine animals eliminate ammonia continuously into the water around them. The fate of this ammonia depends on precisely where it is released. In the tiny spaces between grains of sediment on the bottom, bacteria may go to work, converting ammonia to nitrate. In water filled with growing algae, it may be quickly resorbed and then reassembled into larger molecules with a fresh input of solar energy, starting the cycle anew. Because nitrogen is often in short supply in marine ecosystems, the speed and efficiency of nitrogen recycling may well determine the system's health and productivity.

But this circuit is only part of the many-branched nitrogen cycle. Dead animals and plants contain nitrogen locked up in proteins and amino acids that primary producers cannot absorb. In order to keep the nitrogen cycle moving, large organic molecules must be broken down by bacteria to release their nitrogen in simpler forms. This is an extremely important step; the rate of decomposition and the rapidity with which primary producers assimilate available nitrogen are critical factors in determining the rate at which the nitrogen cycle revolves.

While the earth's organisms scramble for available forms of nitrogen, most of the earth's nitrogen supply floats out of reach above them as nitrogen gas in the atmosphere. Unfortunately for primary producers, the paired atoms in nitrogen gas are held together by an extremely strong chemical bond that no higher plant and no true alga can break apart. Only a few kinds of bacteria, called nitrogen fixers, are able to break the chemical bond and incorporate atmospheric nitrogen into living tissue. The most familiar nitrogen-fixing bacteria live on the roots of such terrestrial plants as peas and soybeans. The most important nitrogen fixers in the sea are photosynthetic bacteria commonly and incorrectly called blue-green algae. Blue-greens, common in shallow-water habitats from the temperate zone to the tropics, pull extra nitrogen out of the atmosphere and inject it into food webs wherever they grow.

Given the intricacy of each nutrient cycle viewed on its own, biologists attempting to understand the workings of all the cycles together face a forbiddingly complex task. The movement of each element through an ecosystem is closely tied to the travels of whatever other elements combine with it in organic compounds. As long as atoms are bound in a single compound, they must travel together through the ecosystem, and protein molecules alone contain a long list of major and minor nutrients, including carbon, nitrogen, phosphorus, sulfur, iron, magnesium, and often many others.

Yet in order for the ecosystem to keep working, every nutrient cycle must revolve at the proper speed. The whole assemblage of cycles works like an infernally complicated set of interlocking cogwheels. The steady one-way flow of energy through the system provides the power to drive the gears in their perpetual revolutions. But as the ecosystem's machinery operates, each wheel must keep turning at its own special rate. The speed of the entire assemblage is controlled by the speed of the slowest wheel—that is, the nutrient that is in shortest supply or the one that moves through the ecosystem most slowly. If any single nutrient cycle is interfered with—if any wheel sticks—the system as a whole grinds to a halt.

In real life, this means that primary productivity may be limited if a single nutrient is in short supply, a phenomenon called nutrient limitation. Ponds and lakes, for example, often contain sufficient nitrogen, iron, magnesium, and several other nutrients to support much higher plant growth rates than are usually found. The primary producers in these ecosystems—and hence the systems as a whole—are stymied by a lack of phosphorus. If phosphorus is suddenly added in large amounts (in the form of phosphate-containing laundry detergents, for example), the phosphorus wheel becomes unstuck, and in no time algae grow prodigiously, covering ponds and rivers and choking out the underlying plant and animal life.

In the sea, on the other hand, phosphorus is often plentiful, and nitrogen is commonly the limiting nutrient. For this reason, the addition of nitrogen-rich sewage and agricultural runoff to coastal waters can greatly increase the productivity of marine systems. In some cases, small-scale release of nutrients from human activities can be beneficial; where domestic sewage from a few seaside cottages leaks slowly into coastal ecosystems such as salt marshes or mangrove swamps, the extra nitrogen may increase primary productivity, enriching the already productive ecosystems. But massive discharges of untreated sewage from coastal towns can encourage the growth of usually rare organisms, can either inhibit the growth of normally important species or cause them to increase out of control, and can thereby seriously upset the local ecological balance.

Adding a single nutrient does not completely free the ecosystem from restraints, of course; it simply allows the machinery to speed up until limited by another wheel. If a superabundance of nitrogen is added to an ecosystem previously limited by it, the nitrogen wheel can speed up; but a new limit will be imposed by the level of phosphorus or potassium or some other nutrient when

Tireless yet selective grazers, sea urchins crop large amounts of algal primary production in shallow water.

its wheel reaches maximum speed. Productivity will increase somewhat, and the system will stabilize at a new steady state. If surpluses of all nutrients are present, the system may be limited by low temperature or insufficient sunlight.

Limited sunlight does, in fact, restrict production in a larger volume of ocean water than any other factor because, although the sea contains a vast store of virtually all essential nutrients, most are effectively locked away. Dead marine organisms and the nutrient-rich solid wastes of free-swimming animals invariably sink away from the surface and out of the photic zone – the only place where primary producers could use the nutrients freed by decomposers. Large quantities of nutrients, therefore, end up in the cold, slow-moving currents that traverse the ocean floor.

While many nutrients sink out of the photic zone, primary producers growing near the surface rapidly absorb the ones that remain. The combination of sinking and photosynthesis thus thoroughly depletes essential plant nutrients in most areas where plants can grow and produces nutrient-rich water on the ocean floor where no plants can make use of them. Luckily for marine organisms, the situation is not as hopeless as it might appear: several important phenomena enable primary producers and nutrients to get together.

The first of these, one that unites widely separated oceanic provinces in completing the nutrient cycle, is a phenomenon called upwelling. In a finite number of places around the world, combinations of winds and ocean currents draw large quantities of surface water sharply away from the edge of a large land mass or relatively shallow bank. Bottom water, complete with its load of nutrients, is pulled up into the photic zone to replace the water thus removed. In places where upwelling occurs, extremely high densities of phytoplankton thrive and grow at rapid rates, providing the basis for a vigorous and productive food web that may include anchovies, herring, lobster, cod, hake, bluefish, tuna, and many other important kinds of seafood. The difference in productivity of the system, even at higher trophic levels, is enormous. Up to 50 percent of the worldwide fish catch comes from upwelling areas, which make up a mere tenth of 1 percent of the total ocean surface.

Upwelling sites are therefore very special places, both to marine organisms and to people who make their living from the sea. One well-known upwelling area is George's Bank off the coast of New England. For centuries, New England fishermen have known that George's Bank is one of the North Atlantic's richest fishing grounds, without knowing why it is so. More recently, both biologists and fishermen have learned that the banks are an important nursery and spawning ground for marine creatures of all kinds, which migrate to the banks from miles away.

The productivity of George's Bank is no longer a mystery; upwelling and mixing currents in the area move nutrients into the photic zone and stir them around in shallow water, providing ideal growing conditions for plant plankton. Although to the untutored eye the water above the banks may look no different from the rest of the storm-tossed North Atlantic, its plankton growth is prodigious and supports abundant fish, bird, and marine mammal life. This special ecological role has earned George's Bank the tenacious support of environmentalists, ecologists, and fishermen. For better or worse, sedimentary rock beneath the bank signals the presence of major oil deposits there, generating strong political pressure from energy producers to develop those resources as quickly as possible. But oil drilling, extraction, loading, and transport are difficult and risky maneuvers in the North Atlantic. Violent winter storms can damage equipment and stymie efforts to clean up oil spills for days or even weeks at a time. It is thus

Behind withdrawn tentacles lie the delicately pleated linings of an anemone's digestive cavity.

a legitimate concern over potential threats to the bank's ecological and commercial value as a food supplier—rather than some obstinate, abstract objection to change—that impels New England states to continue battling for strict regulations to govern development of the area and to keep George's Bank safe.

Another situation that encourages high primary productivity in the sea is created by freshwater runoff from large land masses. As rain water percolates through the soil and travels downstream to the sea, it dissolves and carries a variety of nutrients useful to plant growth. Some of these, naturally, are used by fresh-water algae, but reasonable quantities of nitrogen and other nutrients are carried all the way to the sea. When fresh water meets salt water, the nutrients it carries (such as nitrogen) often complement those present in sea water (such as phosphorus), producing a nutrient mix capable of supporting abundant plant growth. Because fresh water meets the sea in shallow areas where the bottom is within the photic zone, attached algae and higher plants can share this localized nutrient bonanza with phytoplankton.

Both of these phenomena occur mainly or exclusively along the shores of major land masses. Although open oceans are far from being the deserts biologists used to think them, the greatest production of living tissue in the sea almost always occurs near shore or in shallow areas such as George's Bank. The intimate relationship between near-shore areas and fisheries is important to remember, for these areas are especially vulnerable to the many kinds of pollution that can result from human carelessness.

The Manner of Eating

It is no surprise, given the difficulties of obtaining energy and nutrients in the sea, that marine organisms have evolved an extraordinary variety of feeding techniques. Even the briefest look at feeding structures of marine animals will confirm that evolution has left no potential food source untouched. Feeding devices—assembled from skin, scales, bone, and cartilage—often present appearances worthy of creatures from outer space. To nonscientists, these devices evoke images of beauty, serenity, torpor, ferocity, or just plain oddity. To biologists, they tell stories of evolutionary adaptation no less improbable than anything penned by Asimov or Heinlein. Shaped by natural selection, each structure has its own tale to tell.

Even the most mundane marine plant-eaters collect their food in unusual ways. Sea urchins, ranging in size from the diameter of a quarter to the size of a soccer ball, graze tirelessly. Crawling ponderously across the bottom or along kelp stems like aquatic armored tanks, they scrape their plant food from beneath them with a five-part jaw called an Aristotle's lantern. Not to be outdone by sea urchins, certain herbivorous fishes (such as parrotfish) use beak-like teeth and powerful jaws to chew through algae and coral rock with the ease of children nibbling on Ritz crackers.

The techniques marine carnivores use for capturing prey would strain even the most fertile imagination. Goatfish use taste buds on the ends of long, flexible barbels to search in the sand for worms, shrimp, and mussels. Predatory fishes use jaw adaptations to feed in countless specialized ways. Groupers and moray eels use gaping, fang-filled jaws to swallow in a single gulp fishes and

crabs they surprise in the cracks and crevices of the reef. Wrasses and puffers are no less formidable to snails, small crabs, and sea urchins, whose armor is no match for these predators' powerful front teeth. *Priacanthus,* whose ancestors patrolled the coral reef while dinosaurs ruled the land, has kept for eons the simple, open-and-shut mouth of its fossilized kin. *Epibolus,* on the other hand, sports a more recently evolved mouth assembled from a Rube Goldberg–style assortment of natural levers, joints, pulleys, and extra skin. Folded neatly beneath its eyes when at rest, the fish's jaws literally explode into an elongated tube that sucks in and engulfs small prey within milliseconds.

Some carnivores, including lobsters, depend on brute force rather than artifice or cunning to obtain their meals—crushing prey in their claws and then grinding it to a pulp with powerful jaws. Lobsters even have a set of teeth inside their muscular stomachs.

Starfish depend as much on endurance as on strength. Rows of hydraulic-powered tube feet that line the bottoms of their arms attach to hapless clams, mussels, or snails and pull relentlessly in opposing directions. Sooner or later, the prey tires, its shell gaps a bit, and the starfish turns its stomach inside out through its mouth and pours digestive enzymes over the prey. As the prey's tissues begin to disintegrate, the everted stomach surrounds them and draws back into the starfish.

The beautiful members of the phylum *Cnidaria*—the corals, sea anemones, and jellyfish—capture their prey with specialized stinging cells borne in the outer skin of their tentacles. These stinging cells produce structures called nematocysts: poison-filled sacs, each enclosing a tightly coiled, spring-loaded dart. When another animal touches a nematocyst, the dart uncoils explosively, buries itself in the skin of the unfortunate prey, and injects the poison in which it was bathed. To many small swimming animals, nematocyst poison is instantly fatal.

All of these are merely aquatic variations on themes of herbivorous and carnivorous behavior pursued in different ways by land-dwellers. But aquatic animals can exploit food types that simply do not exist in most terrestrial environments. On land, plants, most animals, bacteria, fungi, and all dead organisms rest firmly on the ground or perch on other organisms. Animals that can fly or suspend themselves in the air are large enough and fast enough that they must be pursued and captured individually. The air itself, therefore, carries little in the way of easily obtainable food.

In water, however, floating phytoplankton, swimming zooplankton, and suspended bacteria, fungi, and bits of organic matter are everywhere, and many are incapable of performing escape maneuvers. Near the mouths of rivers, swamps, and marshes, significant concentrations of amino acids and other organic molecules are often dissolved in the water. Surrounded by this floating smorgasbord, marine animals of all sorts have evolved feeding structures that are intriguing and often extremely beautiful.

In availing themselves of the nutrients in plankton, marine species ranging from minuscule worms to gigantic whales have evolved mouthparts, gills, and limbs adapted to filtering the water around them. From the temperate zone to the tropics live sedentary worms that stretch out brightly colored and delicately feathered gills from tubes that shelter the rest of their bodies. These gills, amply supplied with blood vessels for breathing, are also covered with countless tiny hairs and a coating of sticky mucus. When phytoplankton or nutritious bits of larger organisms come in contact with the gills, they are caught in the mucus and funneled by the beating hairs down the spiral groove into the worm's mouth. And with a few variations, this same technique is used by clams,

Colonial anemones paralyze zooplankton with stinging cells in their tentacles; prey are then passed to the polyps' central mouths.

Predators themselves, bristleworms arm themselves with fragile, needle-sharp spines that break off in larger predators' flesh.

The delicate gills of featherduster worms filter small plankton and bits of detritus; they then pass the captured particles to the mouth hidden among the gill bases.

mussels, and a host of other marine species.

Throughout the world ocean, too, such fishes as menhaden and anchovies roam, their gills organized into fine, meshlike nets. These fishes swim with their mouths perpetually open, trapping plankton as water flows over their gills. Every now and then, they close their mouths to swallow collected food before spreading their nets once again.

Even great blue whales and whale sharks, among the largest creatures that have ever lived, are filter feeders, although these giants feed on larger forms of plankton than do the tiny worms. Where their ancestors' teeth might have been, these animals carry huge, brushlike baleen plates that encircle their upper jaws. To feed, these whales fill their mouths with huge quantities of water and plankton, pull their lips in to seal the baleen ring, and use their enormous tongues to force water through the baleen. Once the water has been filtered out, the whales use their tongues to scoop up small shrimp, called krill, and other planktonic organisms caught on the inner baleen surfaces. The tongue then rolls the collected food into a ball and directs it down the gullet.

By filter feeding, baleen whales demonstrate a most efficient technique: harvesting near the base of the ecological pyramid. Animals the size of blue whales—up to 30 meters long and 123 metric tons in weight—need prodigious amounts of food. Were they to feed on large fishes, the total amount of primary production needed to support them would be nearly inconceivable because of the many trophic levels between them and the single-celled algae on which oceanic food webs are based. Good-sized fishes such as tuna and marlin, for example, are usually about five trophic levels away from primary producers, and thus require a primary producer biomass up to 100,000 times their own weight to support them. The number 100,000 is 10^5; as noted earlier, each trophic level can contain only about a tenth of the biomass of the level preceding it, and this creates a geometric progression that quickly becomes staggering as additional trophic levels are interposed between the feeder and the primary producer. If whales ate only marlin and tuna, setting them six levels away from the phytoplankton, the oceans could not support very many of them. By feeding instead on small animals only one step away from primary producers, large filter feeders bypass a number of wasteful intermediate trophic levels.

Another food source of major importance in marine systems is a diverse assemblage of precipitated proteins, decaying bits of animal and plant remains, and fecal pellets containing partially digested food. This eclectic collection of drifting debris is euphemistically called detritus, a term used to describe almost any small organic particle that is no longer alive. No matter how questionable their source, most forms of detritus can provide energy and nutrients to animals equipped to collect them.

While detritus does occur on land, it normally settles on and—in the absence of tornados and windstorms—often becomes part of the soil surface. In the ocean, the nearly neutrally buoyant particles can float and drift in currents for many days and over many miles before sinking to the bottom. Recent studies have shown that many filter-feeding organisms such as worms and corals, previously thought to depend entirely on plankton, obtain a substantial amount of nutrition from detritus.

Detritus feeders, too, can take many forms and utilize varied feeding techniques. Sea cucumbers, now known to be among the most abundant inhabitants of the deep sea floor, have five mucus-covered feeding tentacles, which the animals press against mud and sand as they crawl. Once a tentacle is covered with a mixture of detritus, mud, and sand, the animal pushes the appendage into its mouth, and in the manner of a child licking

Bodies protected by calcareous shells, tubeworms extend their gills to breathe and feed.

Following pages:
Nestled among colonial sea anemones, a solitary barnacle combs plankton from the water with bristle-covered legs.

The giant clam, here surrounded by fire coral, lives on two trophic levels at once: while the clam filters passing plankton, algae living in its mantle photosynthesize.

chocolate cake batter off a finger, sucks off the attached material. Then, like an earthworm, it digests the useful organic material and passes out the sand grains in its feces.

With all these food sources and feeding techniques available, the ecological interrelations of marine organisms are naturally complex. In what they actually eat, marine organisms range from the choosy gourmet to the indiscriminate gourmand. Even in kelp forests and other places where most primary production is accomplished by a single species, some herbivores prefer specific parts of the host plant, some filter plankton from the water, and some dine on smaller attached algae. A whole suite of carnivores choose among the herbivores for prey, often switching from one species to another as they grow larger and as seasons change. In open-water ecosystems, the phytoplankton population can contain dozens of different species, each of which may be the preferred food of several larger zooplanktonic species that vary seasonally in abundance. Add to this the larvae and young of fishes such as herring, which start their lives eating the smallest types of plankton and end up dining on some of the largest, and the complexity of the system becomes apparent.

Clearly, the widely popularized notion of simple, linear food chains – in which big fish eat little fish that eat tiny fish that eat phytoplankton – is a gross oversimplification in most cases. It does describe the first steps of a few simple systems, such as the one involving krill in Antarctica, and it might be useful as a model to help study energy flow through small parts of other ecological systems, but so simple a concept cannot even come close to representing the shifting web of relationships that exists among plants, herbivores, and predators. Rather than forming food chains, these species are interconnected in complex food webs.

In many ecosystems, feeding relationships in food webs contain built-in systems of checks and balances that regulate the population sizes of species in the web and preserve the stability of the whole living system. So complicated and so finely tuned are some of these regulatory mechanisms that some scientists have begun to view whole ecosystems, and even entire groups of ecosystems, as more than just assemblages of independent animals and plants.

For years, biologists have referred to the entire collection of living organisms on our planet as the biosphere, or living globe. The living, breathing, moving biosphere is enormous, multifaceted, and constantly in flux. The relationships among the ecosystems that compose it are too complex to be described by even the most sophisticated of today's computer models. The feeding relationships within ecosystems and the transfer of nutrients and organisms between different ecosystems combine to weave life in the sea into an intricate, living net that stretches over the entire globe.

This network is very real, and its unity exceptionally important. Pluck one supporting rope, and the entire network reverberates. Fray or tangle the net a bit, and it struggles to repair itself; like all living entities, it possesses a substantial ability to recover from disturbance. But tear one portion of the net badly enough – through chronic pollution, overexploitation, or uncontrolled development – and large portions of its fabric may unravel. Because energy and nutrients flow and cycle throughout the network, serious ecological disturbances do not remain localized; they often have far-reaching, even global repercussions. As the human race continues to grow, both in size and in rapacity, it encounters more and more frequently the limits of this network's ability to withstand human disturbances.

The sea cucumber (*top*) collects detritus with mucus-covered tentacles; the sea pen filters both plankton and detritus from passing currents.

A dock piling in New England carries filter-feeding barnacles, hydroids, sponges, a carnivorous starfish, and both macro- and microscopic algae.

Flexible, water-filled suction cups, the tube feet of starfish and their relatives provide power for both locomotion and feeding.

Levers of bone and hinges of cartilage extend this grouper's mouth into a tube, enabling it to engulf and trap its bite-sized prey within a fraction of a second.

This moray's sharp, inward-facing teeth pierce and trap the small, reef-dwelling crabs and shrimp on which it feeds.

This pufferfish's powerful jaws crush the shells of small snails and crustaceans.

Adapted to a bottom-dwelling life, this flounder has both eyes on one side of its head; it seizes small invertebrates with jaws that have retained their original orientation.

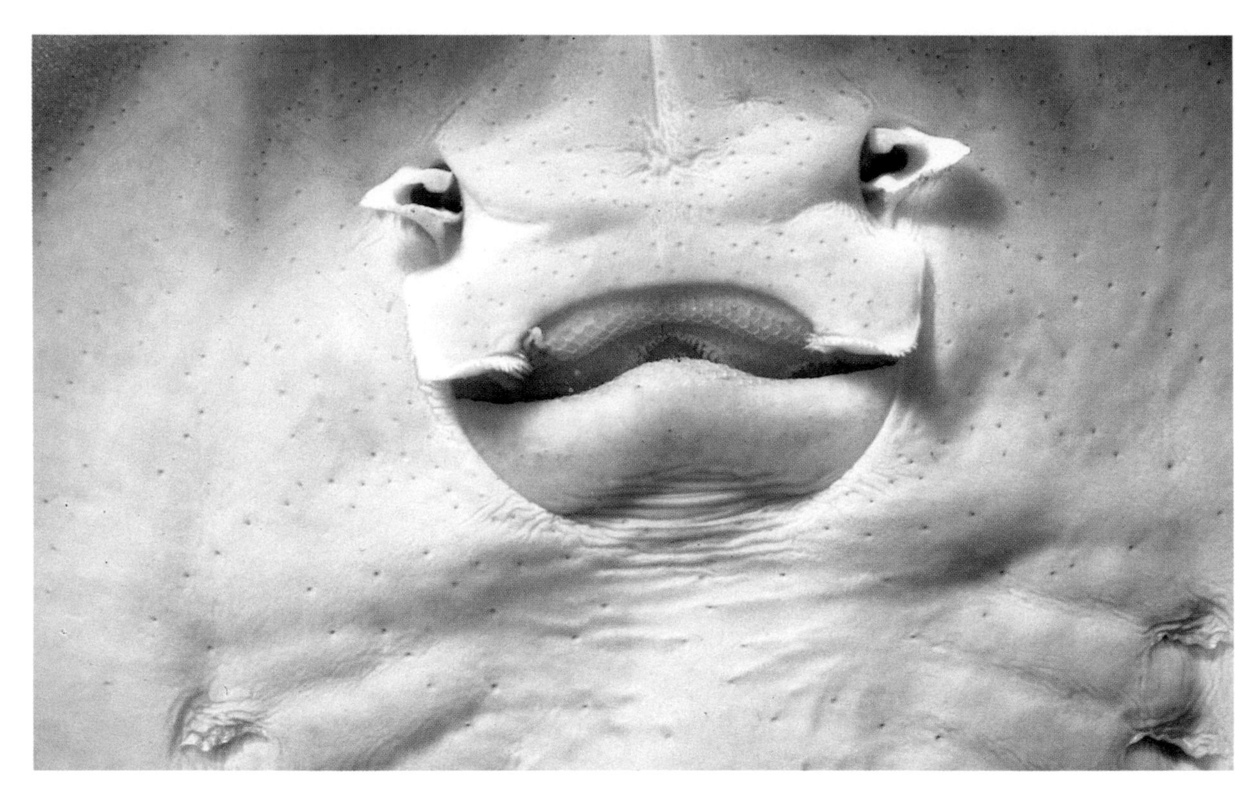

Guided to prey by nostrils flanking its mouth, this skate crushes snails and clams with batteries of tough, rounded teeth.

Right:
Unicorn fish—like their relatives, tangs and surgeonfish—use small but conveniently located mouths to crop algae attached to reef surfaces.

Survival and Reproduction

A mated pair of butterflyfish patrols its territory on a Red Sea reef.

Take a close look at any organism, marine or otherwise, and you will see that it appears to have only two fundamental things on whatever passes for its mind: food and sex. Not only animals' behaviors but their major body structures as well are organized to gather food, to produce reproductive cells, and to use those reproductive cells. Other obvious necessities, such as defenses against predators, simply enhance the organism's ability to survive so that it can eat and reproduce.

Exploring the reason for this simplicity of purpose and structure touches directly upon the keystone of modern biological science. Animal and plant species alive today are those whose forebears have successfully run the evolutionary gauntlet called "survival of the fittest." Nearly everyone has heard this phrase, but few people have an accurate idea of what either "survival" or "fittest" really means in an evolutionary context. The simplest way to come to an understanding of the phrase is to backtrack and follow Darwin's line of reasoning as he worked out the theory of natural selection.

Darwin's thinking was heavily influenced by the writings of the English economist Thomas Malthus. Malthus observed that the human population of his day was continuously growing because babies were being born at a faster rate than people were dying. If the population increased without end, Malthus reasoned, sooner or later it would run out of space and food. The only external checks Malthus saw on such endless growth of human populations were food shortages, plagues, and wars. These less-than-optimistic observations about human population growth became known as the Malthusian Doctrine.

Darwin realized that the same issues of population growth addressed by the Malthusian Doctrine must arise with even more force in populations of other organisms because most animals and plants produce far more offspring than humans do. A mature female starfish, for example, can produce more than one million eggs each time she spawns. If all the offspring of one such starfish survived for one generation, they would completely fill the area in which they lived. If each of these offspring had as many offspring as its parents did, and if all of those offspring survived, there would scarcely be room to squeeze them onto the entire west coast of the Americas. If all the offspring of the third generation survived, and so on for only about fifteen generations, they would not only overrun the entire planet, they would be more numerous than the estimated number of electrons in the entire visible universe!

Obviously, the world is not overrun with starfish, nor is it buried in mussels, crabs, or flounders. Most starfish larvae die during their first few weeks of life. Most survivors of round one die before reaching sexual maturity. But Darwin puzzled over the factors that determine which individuals within a population survive and reproduce.

Darwin observed that organisms in nature are engaged in a constant struggle for survival. They must live through cold, heat, and other unfavorable environmental conditions. They must compete with one another—and especially with others most like themselves—for suitable food and living space. As a matter of course, many animals must kill and eat other animals to survive. From his own experience in animal breeding, Darwin knew that every individual organism is slightly different from all other members of its species (although he did not know why). He also knew that some of this variation among individuals can be passed on from one generation to the next (although he did not understand how).

He then reasoned that, in the struggle for existence, individuals would tend to survive that possessed the particular characteristics that best enabled them to deal with conditions in their local

Delivered prematurely from its mother's reproductive tract, a developing shark embryo exhibits the yolk sac that normally supplies its nutritional needs until birth.

environment. Individuals lacking the right characteristics would either die before they had a chance to reproduce or leave fewer offspring if they did reproduce. As generations passed, organisms with characteristics best suited for survival and reproduction would succeed in passing those characteristics on to their offspring, and their line would become more and more common. This process, the differential survival and reproduction of organisms with different heritable characteristics, Darwin called natural selection.

Today, the science of genetics provides a great deal more information about the processes Darwin observed. Heritable characteristics are controlled by genes located on chromosomes. The process of sexual reproduction makes copies of the parents' chromosomes, shuffles them around like a deck of playing cards, and deals them out to the eggs and sperm. Each time the chromosomes are shuffled and dealt, they combine in different ways, and this provides one source of variation among offspring.

Geneticists have also discovered that other sources of variation among individuals arise when genes are altered. Mistakes in the copy can occur when chromosomes are duplicated. Genes can hop around from one place to another on the chromosomes. Radiation, certain chemicals, and some viruses can also cause genetic changes. All of these genetic alterations, which can produce heritable results in the next generation, are called mutations.

From the standpoint of evolution, mutations are random events. They have no plan or purpose; they simply happen, without regard to whether they will ultimately be useful, useless, or harmful to the organism. Mutations do not occur because an animal needs or wants to evolve. Neither animals nor their environments can cause a specific mutation, nor can they prevent one. (Humans can increase the rate of mutations by exposing cells to radiation or chemicals, but even so they cannot control just what mutations occur.)

Genetics has produced a modern definition of fitness, too: an organism's fitness is measured by the number of copies of its genes it contributes to the next generation. The old adage "a hen is just an egg's way of making another egg" has been updated to "an organism is just a gene's way of producing another set of genes."

Some evolutionary biologists who followed Darwin popularized natural selection as being far closer to omnipotent in directing evolutionary change than Darwin himself believed. They also interjected into their version of Darwinian thought a positivistic and anthropocentric bias drawn from Western philosophy that was actually antithetic to Darwin's own hypotheses. Evolution, these biologists argued, produced constant progress toward perfection in the living world. Older, less well-adapted, or "primitive" forms are replaced by others blessed with better design. In this master plan of progress toward perfection, of course, *Homo sapiens* in its modern, Western form was placed at the apex of all evolutionary development.

Biologists thinking in this manner encumber students of evolutionary theory with two related misconceptions: that all animals are perfectly designed to do whatever it is they need to do; and that living species are somehow superior to their extinct predecessors, having progressed in excellence during the course of evolution.

The first of these ideas, which attempts to see each anatomical, physiological, and behavioral characteristic of each organism as "optimal" for performance of its required task, has been dubbed the Panglossian paradigm by Harvard biologists Steven Jay Gould and Richard Lewontin. Gould and Lewontin noted a great similarity between this view of biology and the world view of Candide's tutor, Doctor Pangloss, who was given to such pronouncements as: "Observe that noses were made to carry spectacles; and so we have spectacles. Legs were visibly instituted to be breeched, and we have breeches." Hand in hand with the idea of perfection and optimization in structure goes the conviction that every characteristic of every organism necessarily evolved to serve a specific purpose, regardless of how convoluted its structure might be.

The misguided idea that ancient classes and orders of extinct or living organisms are evolutionarily inferior to more recently evolved forms permeates biology textbooks from high school onward. Witness the description of invertebrates as "lower animals" and of fishes as "lower vertebrates."

More and more evolutionary biologists, however, are simultaneously returning to Darwin's original ideas and adding new concepts and mathematical insights. First, they reemphasize Darwin's original contention that natural selection has no guiding purpose, although the extraordinary feats of biological engineering that surround us make it tempting to posit one. Natural selection, operating on individuals of a species and their genetic makeup, simply selects for characteristics that contribute to survival in local environments that are rarely stable and often stressful.

Alongside fitness-related changes influenced by natural selection, however, a significant proportion of evolutionary change is generated simply by chance and has little to do with the selection pressure. Mutations that are not necessarily the "best" for an organism, but that do not decrease its fitness either, occur and persist in populations through random genetic drift without causing problems for the organisms that carry them.

Rather than viewing evolution as a forced march down a narrow corridor, directed by the all-powerful force of natural selection, these biologists see evolutionary change as more of a random walk along a network of aimless paths through a very thick woods. Evolutionary change has neither a goal to get from point A to point B nor a plan for accomplishing any such goal. Depending on the neck of the woods they happen to find themselves in, species can blunder along the paths, changing as they go, or they can dawdle. Some continue to exist, while others stumble off into extinction.

According to this view, many differences among present-day organisms simply resulted from chance mutations creating different solutions to the same problems. If two species do things differently, one way is not necessarily better than the other; each species may simply have acquired a different workable alternative along its path of random genetic change. More and more biologists feel that random events have played a decisive role in creating the diversity of organisms around us. If every characteristic of every organism had to be shaped by natural selection, if a single, optimal way existed of performing a required task, and if chance played no part in the process, most organisms would perform important functions in essentially the same way. If, on the other hand, each solution evolved by a species represented just one of a set of workable alternatives, species would be under no imperative to improve their performance unless environmental changes forced them to.

Yet this basically haphazard process has produced innumerable organisms that amaze those who study them. Words and phrases such as "clever," "ingenious," and "well-engineered" pepper the biological literature, bearing witness to the admiration plant and animal structures and functions evoke among scientists. These words, however, should not be taken to imply a human sense of purpose behind evolutionary change. Natural selection operates with no relation to whether an adaptation might be deemed "clever" or "inventive." Such terms—especially in popular books and articles—are meant either to describe human reac-

Egg ribbons of Spanish Dancer nudibranchs contain hundreds, even thousands of eggs.

Depositing its eggs among algae and filter feeders, this nudibranch will summarily abandon its brood as soon as the job is done.

tions to the remarkable products of evolution, or to serve as scientists' shorthand for more accurate and objective, but cumbersome prose.

The marine environment is filled with species whose habits and histories are far more amenable to this view of evolution as undirected and pragmatic than to the popularized notions of programmed, goal-oriented change. Along the Atlantic coast, fishes that evolved as recently as humans live side by side with horseshoe crabs that are nearly indistinguishable from fossilized ancestors over 200 million years old. On tropical coral reefs, crabs and fishes that have evolved since the recent ice ages live side by side with feathery crinoids that are practically identical to relatives extinct for 500 million years.

Sharks, identified as "the least complex vertebrates" by the most widely used high school biology text in the United States—as well as by several college texts—are so labeled because they have skeletons made of cartilage rather than of bone and because they first appeared in the fossil record about 400 million years ago. Yet these "simple" fishes evolved and retain many unique and thoroughly workable features that serve them as well now as when the dinosaurs flourished. For example, sharks need never face tooth decay or wear: their gums produce rows of teeth in endless succession. As older teeth become worn or damaged, they simply fall out, to be replaced in short order by the next in line.

In addition to sharp eyesight and an extremely acute sense of smell, some of these creatures also possess an electrodetection sense completely unknown among "higher" organisms. Pores and canals set in rows along their snouts contain sensory cells that detect the minute electric currents generated whenever prey move and breathe. These same electrosensitive cells also detect the tiny electric currents created as ocean currents cross the earth's magnetic field, enabling sharks to navigate accurately over long distances.

Other marine organisms offer innumerable opportunities to examine the diversity that the random paths of evolutionary success have generated in different organisms. Reproduction is an excellent function to investigate for such differences among animals, for here, as with any critically important biological function, evolution has produced a wide variety of working solutions. Marine animals, famous among biologists for evolutionary variety, go about sexual reproduction in an unusually large number of ways.

Some animals, including many butterflyfish, are monogamous; they mate for life, defend territories as couples, and care for their young as a pair. Others, including most schooling species, are promiscuous; traveling in swarms, they mate as a group, shedding eggs and sperm into the water column with no regard for which sperm fertilize which eggs. Still other species, including the wrasse *Labroides,* live in harem groups in which a single male shares a territory with five or six females. The resident male chases all other males away and mates with one or another member of his harem every day.

The actual process of mating also takes innumerable forms. The male horseshoe crab, for example, grabs onto the female, who then carts him around on the sand until the eggs are laid and fertilized. Another example is the female crab or lobster, who can mate only after molting. Covered only by a soft, thin shell, she is at the mercy of the male, who may either fertilize or eat her.

Most marine animals lay eggs and fertilize them outside their bodies, but others, including many sharks, fertilize eggs internally: the males use specially shaped fins to deliver sperm into the reproductive system of the females. Some species that practice internal fertilization lay eggs that develop externally; others nourish developing embryos internally and bear their young alive.

Its snout equipped with sensitive nostrils and hundreds of electrodetecting pores, this sand tiger shark rips chunks of flesh from its prey with rows of replaceable teeth.

Male horseshoe crabs attach to their consorts, with
whom they travel until eggs are laid.

Horseshoe crabs are common in New England
waters, testimony to the success of the life-history
strategy the species evolved millions of years ago.

In a head-to-tail nuptial embrace, hermaphroditic nudibranchs exchange sperm.

Numerous fish and invertebrate species can and do change sex—whenever conditions warrant—with astonishing speed. An example is the tropical reef fish *Anthias squamipinnis.* This brightly colored basslet swarms over coral formations in schools that contain hundreds of individuals. Males and females look so different that most casual divers assume them to be two different species. Females—which usually outnumber males by about ten to one—are a striking orange-red all over, while males are dramatically patterned in violet, purple, and white. Even more intriguing is research showing that virtually all *A. squamipinnis* eggs hatch into females. As the fish mature, they join the mostly female school, where they enter a pecking order in which larger females dominate smaller ones. If a male is removed from the school, the largest or "alpha" female changes into a fully functional male in less than two weeks. If nine males are removed, the largest nine females change sex. Just how the right number of female fish are triggered to become males is still a mystery.

But even this species is outdone in sexual gymnastics. Some animals are synchronous hermaphrodites, capable of functioning as males and females at the same time. Although they rarely if ever self-fertilize, some species do engage in rather spectacular displays of group sex. Certain sea basses mate in pairs, exchanging male and female roles as often as four times during a single mating. Each member of the pair alternately produces eggs to be fertilized and sperm to fertilize its partner's eggs. Several hermaphroditic nudibranch species mate in long "daisy chains" in which each animal delivers sperm to another individual while at the same time being fertilized by a third.

Once eggs are laid, of course, the matter of parenting arises. Nudibranchs put all of their reproductive energy into producing hundreds, or even thousands of eggs in a delicate ribbon. The embryos are summarily abandoned by both parents, however, immediately after they are laid and fertilized. Unlike these marine snails, many female crustaceans do not desert their eggs immediately. Instead, they lay fewer eggs at each mating and carry their young in a brood pouch or under their abdomens until they hatch. Even so, the larvae of these species are left to fend for themselves as they float and somersault in the plankton before metamorphosing into adults.

Other animals, such as clownfish, lay still fewer eggs but glue them together in a small clutch and guard them ferociously against all intruders. Brooding clownfish attack much larger animals that venture too close to the nest. Sharks that bear their young alive use an even more protective brooding technique. Instead of producing many small eggs and deserting them, these species carry a dozen or so developing embryos inside and give birth to them only when the young are fully formed and ready to fend for themselves.

These variations in reproductive strategies—just a few of many that exist in the sea—all bear witness to the unpredictable and phenomenally creative process of evolution. At the same time, they illustrate the impossibility of satisfactorily assigning causality in the evolutionary process or identifying a particular strategy as "best" in a given situation.

For example, *A. squamipinnis* and other fishes that start off as females and later change into males tend to live in groups. Large males either control these groups or vigorously defend special spawning territories, a trait that enables them to reproduce frequently while preventing small males from mating at all. Small females, in contrast, have plenty of opportunities to reproduce. Because any characteristic that allows an organism to leave posterity more copies of its genes is favored by natural selection, any individual that could start mating early in life as a female and then change sex when

Born female, this *Anthias* has assumed the colors, behavior, and reproductive functions of a male.

Following pages:
A paralyzed butterflyfish disappears down the gullet of a large sea anemone.

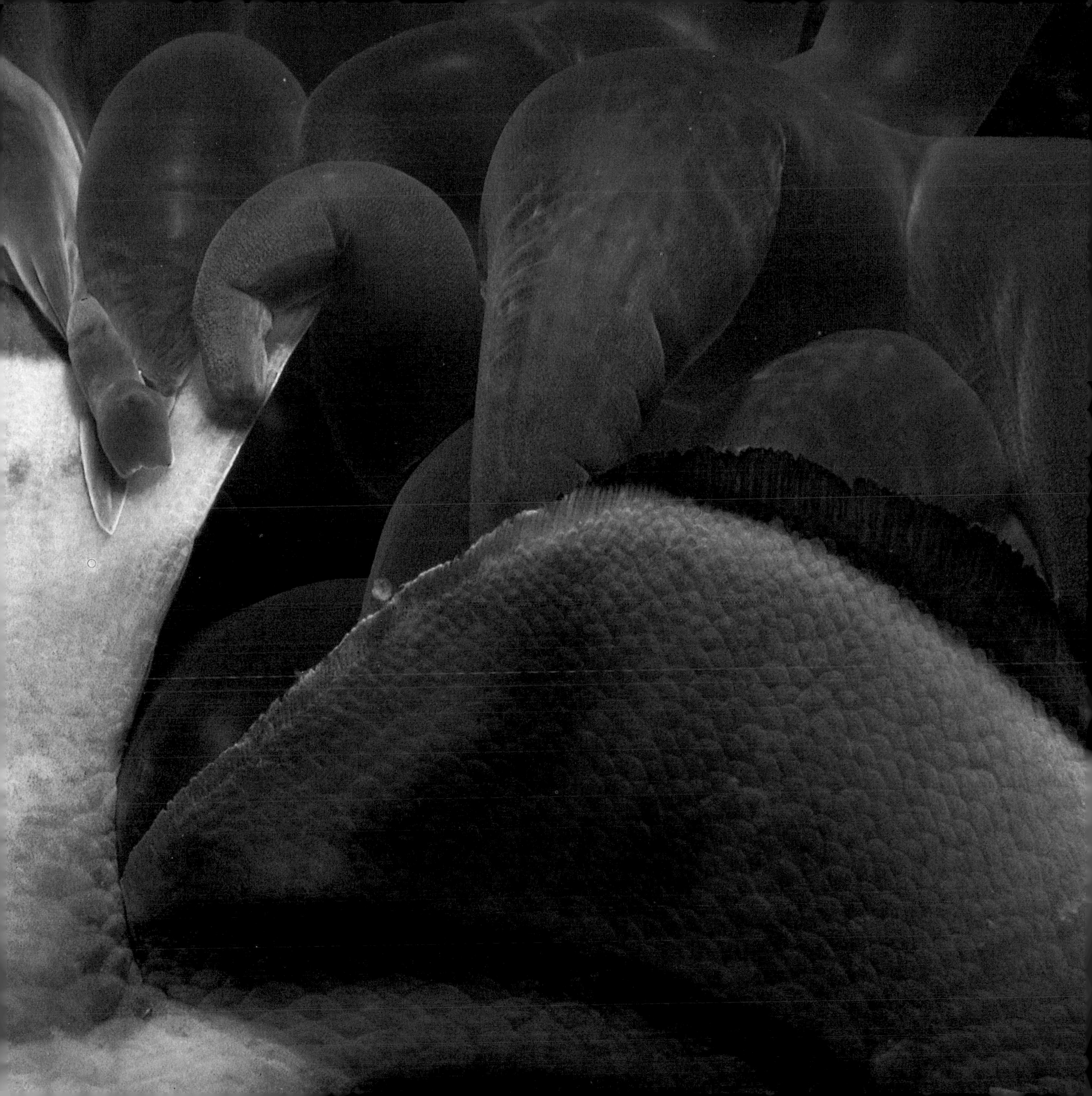

large enough to mate successfully as a male would gain an enormous advantage in fitness under the circumstances of group life.

Clownfish, on the other hand, live in mated pairs for long periods and stay close to a home territory where they mate and lay eggs. Because eggs take significantly more energy to produce than sperm, pairs in which the female is larger can produce the maximum number of fertilized eggs. In fish with this sort of social system, individuals are usually born males and change into females as they grow larger.

Although correlations such as these hold up for many species, the chicken-and-egg question still remains. Which came first—the mating system or the social structure? Both are inexorably linked to species fitness, and both probably evolved together, in concert not only with each other but with the environment around them.

Herein lies the true wonder of the evolutionary process; it presents seemingly endless opportunities for variability. Each species, standing in the middle of an environmental woods, faces not just one, but a variety of potential evolutionary paths, none of which it can wittingly choose, and none of which is intrinsically better than any other. Moreover, the woods is by no means a stable place; each species is surrounded by its predators, its prey, and all the other organisms that constitute its biological environment. The whole assemblage becomes a continually shifting kaleidoscope of changes and counterchanges. Predators evolve new techniques for catching prey at the same time that prey evolve new ways of avoiding them. Social systems, mating behavior, and parental care evolve in concert, as the species wends its way along the ever-shifting evolutionary paths that open and close before it.

Small wonder that the great evolutionary thinkers of our time—Theodozius Dobzhansky, Ernst Mayr, and Julian Huxley among them—have stood in awe of the creativity of the evolutionary process and the diversity of its products. Dobzhansky, in fact, compared natural selection to a composer—a metaphor that, within limits, fits reasonably well. For just as a composer's historical milieu and state of mind can influence the form of a composition, the physical nature of environments and the existing, living network of organisms can guide the random notes of genetic variation into a coherent course of evolution.

The grace of these sea anemone tentacles belies the danger their stinging cells pose to small fishes.

Where the Sea Meets the Land

Rivers and Sea

A green-backed heron pauses between feeding bouts in the Everglades.

Following pages:
Spartina alterniflora covers a flooded mud flat in northern Massachusetts.

The sea breeze over the Cape Cod sand flats carries both sounds and odors of life. Ebb tide abandons rocks, mussels, mud, and buried clams with gurgles, pops, and a nearly inaudible sound like that of crinkling wax paper. Sand bars divide the bright blue waters of the bay with free-form shapes of brown and gray, punctuated by the black and white bodies of terns and seagulls at rest. Behind the retreating ocean, a steady stream of fresh water drains from the tidal river, whose currents tug and pull with the waves, shaping banks composed of water-borne sediments. Here, as in all other places called estuaries, fresh-water runoff from the land mixes with sea water, creating a habitat that is highly productive and strikingly beautiful.

Shoreline and shallows bustle with activity. Sandpipers pick their way through the spray of retreating waves. Seagulls peck among the rocks for worms, crabs, and shellfish. Gulls in possession of clams or mussels that they cannot pry open carry the bivalves high in the air and crack them open by dropping them on the rocks below.

People who make their living by harvesting marine life in shallow waters know well the life that crowds their local estuaries. Fishermen might not understand why estuaries are so productive, but they have long known what good fishing grounds they are. Centuries ago, low tide found Native Americans and early settlers combing New England's tidal marshes, islanders across the South Pacific fishing in their flooded mangrove forests, and Dutch lowlanders harvesting the Oosterschelde. At the turn of the tides, nets were set and lines cast into bays, shallow inlets, and tidal creeks alive with fishes riding the currents. From tiny minnows to the full-sized game fish that feed on them, from herring and alewives to stripers, mullet, tarpon, milkfish, and bonefish, the list of marine life is as long as the number and variety of estuarine locations.

Part of the reason for the great productivity of estuarine ecosystems lies in the mixing of fresh and salt water. Fresh-water runoff from the land—especially but not exclusively in inhabited areas—often contains nitrogen and silica, but may lack phosphorus. Water near the surface in the open sea often carries considerable phosphorus, but tends to be nitrogen- and silica-deficient. Each water type thus has limits on the plant growth it can support, because each lacks sufficient quantities of one or more essential nutrients. When combined in shallow, sunlit, estuarine waters with a strong mixing current and ample sunlight, their nutrients complement each other and encourage high levels of primary production.

Once nutrients have entered estuaries, several factors keep them there in forms that both animals and plants can use. Rivers drop their loads of silt

and clay particles as they mix with tidal currents and meet the sea. These tiny sedimentary particles readily adsorb nutrients from the water around them. Once adsorbed, nutrients in the sediment can be released to primary producers if the nutrient concentration in the surrounding water drops significantly and if other chemical conditions are right. Sediments thus help buffer nutrients in the estuary, picking them up when they are present in excess and releasing them when they are in short supply.

The normal activities of many estuarine animals also help keep nutrients in the area. When phytoplankton grow densely in the estuary, filter feeders such as clams and mussels remove far more cells from the water than they are able to digest. Many of these plant cells pass through their guts undigested, to be excreted in large quantities as semisolid particles called pseudofeces. Because pseudofeces are large and can readily be manipulated by other organisms, and because they contain substantial quantities of useful nutrients, the normal feeding habits of clams and mussels effectively package and store plant food for use by worms, shrimp, crabs, lobsters, and other bottom feeders.

Despite the abundance of nutrients, however, life is hard in these deceptively peaceful-looking ecosystems. Fresh water meeting the sea head-on generates a salinity gradient from full-strength salt water on the ocean side to completely fresh water on the inland margin. Tides alternately flood and expose large areas and constantly alter levels of salinity throughout bays and tidal creeks. Storm waves pound the shoreline, and rainstorms generate sudden surges in fresh-water flow.

For these reasons, although estuaries teem with life, their cast of characters is rather small. Many estuaries are prime examples of habitats where a physically rigorous environment restricts the number of resident species to a hardy few. But because these conditions mean that fewer species live in estuaries than in either fresh-water or marine ecosystems nearby, the physical difficulties of estuarine life are balanced by relaxed competition among different species for food and space. This in turn allows many estuarine species to be ecological generalists capable of feeding on a variety of foods, depending on which is most abundant.

Species that can tolerate salinity and temperature changes and occasional stranding at low tide, therefore, can exploit estuaries' high productivity, multiplying rapidly and growing into populations that contain enormous numbers of individuals. The rich waters of Chesapeake Bay, for example, harbor staggering populations of both oysters and blue crabs, which—except in polluted or overfished areas—can virtually carpet the bottom. Along Long Island's coast, Great South Bay nurtures enough clams to support the largest single fishery in all of New York State. Both of these cases typify estuarine populations: there are very high numbers of individuals belonging to relatively few species, and there is a predominance of animals whose ecological requirements and tolerances are relatively broad.

To people interested in marine ecology, estuaries and the productive plant and animal communities they contain provide perfect points of entry into the maritime realm. As hybrid ecosystems containing both terrestrial and marine characteristics, estuarine habitats look refreshingly familiar from a distance. Temperate-zone salt marshes, their above-water areas blanketed with lush, green grasses, resemble fertile fields and grasslands. Tropical mangrove forests, cloaked in bushes and trees from 2 meters to over 10 meters high, evoke images of dense woodlands.

Beneath their apparent similarity to terrestrial ecosystems, however, lie food webs and nutrient cycles different from those of any terrestrial sys-

Black ducks and other waterfowl find both food and shelter in northern marshes.

tem—cycles that tie together the systems' above-water plant growth and their marine life hidden beneath the water's surface. Different in their particulars, mangrove and salt-marsh systems the world over are nonetheless fundamentally similar in ecological organization and importance to adjacent ecosystems. Marshes of one sort or another grow from the edge of the Arctic Ocean, along the Atlantic coast of Canada, through New England, and down the United States coast as far south as Georgia, as well as in England, Australia, and New Zealand. American salt marshes near Falmouth, Massachusetts, house many of the same species and numerous species from the same genera that inhabit marshes near the mouth of the river Fal in England. Similarly, mangrove forests in the Caribbean and South America house collections of flora and fauna that are equivalent to mangrove areas of the South Pacific. From the importance of estuaries in life cycles of commercially valuable fishes to humanity's need for transportation centers near protected harbors to the worldwide association of marshes and swamps with nuisance animals such as mosquitoes, estuaries and their ecology are intimately tied to the lives of human beings along the ocean's edge.

Temperate Grasslands of the Sea

Banking low over waving marsh-grass fields, a flock of Canada geese brakes and lands gracefully to join scores of ducks resting and feeding during their long fall migration. Ducks, diving deeply in bays and ponds, and geese, pecking at shallow-water vegetation in the meandering creeks that drain the marsh, seem driven by insatiable appetites. At low tide, snails plow ponderously over marsh mud and plant matter, scraping a trail through algae on the marsh surface and climbing grass stems to graze on the thin layer of algae growing there. Fiddler crabs, whose burrows honeycomb the creek banks, thump their claws on the ground and disappear at the slightest disturbance, cautiously emerging a few moments later to eye a quiet observer.

Just beneath the water at the tide's lowest ebb, an army of hermit crabs, clad in once-vacant snail shells commandeered for shelter, ambles along, grabbing here at algae, there at tiny shrimp. Where partially exposed rocks and logs break the sediment surface, a carpet of delicate green, red, and brown algal fronds grows. Beneath the dried upper layer lives an entire community sheltered from sun and open air in the folds of the algal mat. At an intruder's approach inch-long crabs scurry for cover under rocks, quarter-inch amphipod shrimp disappear into the sand, worms snap back into their burrows and tubes, and snails retreat hastily into their shells. Wherever the tide exposes sand or mud flats, flocks of small wading birds forage among clams and mussels and scrutinize algae-covered sediments for tasty animal morsels. Beneath the water's surface, in creeks and bays still flooded at low tide, submerged sea grasses and algae together house and feed yet another entire living assemblage.

The marshes these organisms live in are dynamic systems, old by human standards but quite young on the geologic time scale. The oldest of the present-day salt marshes came into existence as the last glaciers retreated around 15,000 years ago; others date back only 5,000 years. Melting ice raised the sea level around the globe, flooding sloping coastlines and drowning river mouths to create estuarine bays. Over the centuries, winds,

Roots and recumbent stems of dune plants stabilize sand and accumulate organic matter on barrier beaches.

waves, and tides sculpted river-borne sediments and coastal sands into large sand bars that stretched across the bays' entrances. Whenever sand on the shore piled up above the high-tide line, the rugged beach grass *Ammophila* colonized the shifting dunes, holding them in place.

If the sedimentary processes continued, sand bars grew into barrier beaches: long, exposed fingers of sand, dunes, and beach plants that slowly restricted the passage of water between river mouth and sea. From Massachusetts' Plum Island to New York's Fire Island, down the coast to the great estuaries of Delaware and Maryland, and still farther south to the famed sea islands of Georgia, the barrier beaches grew along the Atlantic coast of the United States. The same processes, working in slightly different ways, created similar harbors in coastal European areas.

As the barrier beaches steadily closed the passageways between sheltered bays and the open sea, they forced tides and wind-powered ocean currents to slow down. As each incoming tide lost speed on entering the bay, it lost the energy that empowered it to carry sediment. Dropping whatever sand and silt it carried, each tide added a bit more sediment to the bay floor. When the accumulating sediments had risen a few inches above the midway point between high and low tides, the seeds of the marsh grass *Spartina alterniflora* germinated in the mud beneath the brackish estuarine water. Once this grass established a foothold, the life of the bay was forever altered.

Across the mud and sand flats of the protected estuary, *S. alterniflora* grew quickly, sending out roots and underground stems that knotted together into a tough, dense mat. An aggressive colonizer and invader of new habitats, like many other marsh organisms, *S. alterniflora* crept steadily over the shifting mud, creating a spreading blanket of turf that held sediments beneath it firmly in place. The grass's tall, stiff leaves stretched 1 to 2 meters above the roots, producing a resilient barrier to water flow that further interfered with tidal currents and encouraged still more rapid deposition of sediment.

Like all perennial grasses, *S. alterniflora* produces new roots and stems each growing season from its overwintering underground rhizomes. Every spring, bright green shoots emerge from the stubble of the previous year's growth. Throughout the summer, leaves stretch upward, to be joined early in the fall by long stalks of inconspicuous flowers that mature into autumn's golden seed stalks. Meanwhile, an equally vigorous surge of root growth radiates into the mat below. Building on the combination of previous years' turf and recently deposited sediment, each season's crop raises the marsh surface, keeping pace with slowly rising sea levels. Beneath the actively growing surface layer, old roots, rhizomes, and sediments combine to form salt-marsh peat, whose depth testifies to the relentless upward growth of the marsh.

In many salt marshes today, the steadily rising *S. alterniflora* turf rises close to the reach of most high tides. Here, sediment deposition is slowed, rain and fresh-water runoff lower salinity, and other plants move in. The most successful invader on this high marsh is another member of the same genus, *Spartina patens*, commonly called salt-marsh hay, whose thin leaves and delicate flower stalks contrast markedly with the vigorous, upright *S. alterniflora*. Growing only 30 to 60 centimeters high, and contributing to marsh height much more slowly, *S. patens* spreads across much of the marsh surface in many of the older New England marshes and in marshes in southern coastal states.

An established marsh creates a remarkable ecosystem whose health and well-being are intimately tied to the health of adjacent terrestrial and marine systems. When nearby marine ecosystems are free of dredging and filling operations and when the terrestrial habitats upstream are not changed too drastically by development and agriculture, *Spartina* grasses stabilize the sediments on which they grow and provide much-needed constancy in a shore zone notoriously vulnerable to capricious and destructive coastal storms.

This physical stability enables a few tolerant species of sedentary primary producers to settle in and around the *Spartina* grasses. Nitrogen-fixing blue-green bacteria can grow wherever sufficient sunlight penetrates through grassy cover to the sediment surface. Single-celled algae called diatoms grow on and between sand grains, actually migrating toward or away from the surface to control their exposure to sunlight. Other algae coat *Spartina* stems and grow upward along their leaves.

Shallow, continually flooded salt-water ponds, embayments, and tidal channels harbor communities of one or more of the few totally aquatic higher plants, *Ruppia* (widgeon grass), *Zostera* (eelgrass), and *Thalassia* (turtle grass). Although commonly called sea grasses, these genera are not at all related to the terrestrial plants of the same name. They are an odd lot whose ancestors left the land for a marine existence millions of years ago, probably before the great upheaval of the late Cretaceous extinction, during which the dinosaurs and many sea creatures died out.

Sea-grass meadows are highly productive ecosystems, rivaling even cultivated tropical plantations. Some species of sea grasses can absorb nutrients through either leaves or roots, allowing them to draw upon whichever nutrient source is richer at a given time. Sea grasses are even able to move nutrients around in significant quantities; they can take up ammonia and phosphorus from sediments through their roots, pump them up into their leaves, and send them out into the surrounding water.

Perhaps because their leaf surfaces can supply nutrients in this way, sea grasses are always covered by a complete community of epiphytes—single-celled and multicelled algae, small filter feeders, and hordes of bacteria that grow right over their leaves. The primary producers in this miniature community, along with the leaves of the sea grasses themselves, extract nutrients from river and sea water and, with the help of abundant sunlight, transform them into living tissue. Some researchers assert that sea-grass epiphytes actually provide more useful nutrients to grazing animals than do the grasses themselves. Most animals that eat sea grasses, with the possible exception of manatees in tropical and subtropical areas, have no enzymes of their own for digesting the cellulose that makes up so much of the leaves. Some animals host microorganisms in their guts that produce cellulose-digesting enzymes, but much of the cellulose eaten still remains unaffected during its passage through the gut. As a result, these animals swallow huge quantities of sea grass, digest primarily the organisms growing on the blades, and excrete detritus in large quantities.

Sea grasses often grow in submarine meadows that are lush enough to augment substantially—and, in the opinion of some researchers, even to exceed—the contributions of exposed marsh plants to estuaries' productivity and habitat value. Between their leaves and their epiphytic flora and fauna, sea grasses constitute an important food source for many animals. Sea grasses also help stabilize sediments below the low-tide line where neither marsh grasses nor mangrove trees can grow. Sea grasses may even contribute to the growth of phytoplankton around them by pumping nutrients from sediments into the water column.

In the terrestrial grasslands that marshes superficially resemble, animals of importance to humans—insects, rodents, large grazing mammals,

Pioneering *Ammophila* plants colonize shifting sand, paving the way for less tolerant species.

and birds—graze directly on leaves or seeds. In the above-water parts of the marsh, however, this is not the case, for most important marsh consumers do not eat the living grasses. Instead, *Spartina* roots and shoots die in their own natural rhythms. Older, outer leaves and roots die off over a long period throughout summer and fall. In the north, hard frosts then put a sudden end to the growing season and kill the plants back to their rhizomes.

Dead plant parts are immediately attacked by bacteria and fungi. Secreting enzymes to break down the plants' cellulose skeletons, these microorganisms utilize energy and nutrients not available to most multicellular animals. Decompositional activity helps break plant structures into small pieces, which wash off the creek banks and into the water. There, the detrital particles and the microorganisms associated with them form the base of a food web markedly different from that of any terrestrial grassland.

The constantly growing body of data on marsh systems has begun to reveal a complex and dynamic picture of life in the marsh that emphasizes just how intimately interconnected the processes of life, death, and decay are in this environment. Detritus and the marsh's tiny decomposers are the central players in the drama. Microorganisms alone probably consume nearly half the *Spartina* tissue produced above ground each year. Bacteria and fungi extract many nutrients from detritus, and they grow like living cloaks around countless detritus particles.

As they ingest these particles, the marsh's larger detrital feeders prepare the way for the next cycle of use. By chopping, grinding, and subjecting detritus to their digestive enzymes, detritivores constantly break the particles into smaller and smaller pieces and remove the microscopic decomposers. While doing this, the animals release both dissolved nitrogen and solid nitrogen back into the water. Primary producers quickly resorb much of the dissolved nitrogen and use it to produce new plant tissue. Solid nitrogenous wastes mix with detritus particles and pseudofeces to provide fodder for still another horde of microorganisms and detritus feeders. Again and again this processing occurs, constantly recycling nitrogen, phosphorus, and other nutrients as well. Constant microbial activity and repetitive processing of detritus keep these nutrient wheels spinning rapidly.

Curiously enough, the more research is done on salt marshes, the less agreement there is among researchers as to how nutrients flow out of the system. Part of this confusion probably stems from differences among the particular marshes studied. Some marshes grow near the mouths of substantial rivers, some in sites with little fresh-water input, and some in places with widely varying tidal patterns. Human population growth in major metropolitan areas such as New York City has surrounded some marshes while leaving others relatively untrammeled.

Not surprisingly, the flow of nutrients through these marshes occurs in different patterns, in different forms, and at different speeds. Studies of some marshes indicate significant export of detritus or of dissolved organic matter, transferring both energy and nutrients into adjacent marine ecosystems. Studies on other marshes suggest that under different circumstances nutrients are efficiently recycled within the marsh, leaving relatively little to be carried away by waves and currents.

Even with all the nutrients that make them so productive, however, many marshes are clearly nutrient-limited. Experiments on Cape Cod have shown that many productive New England marshes could grow still more lushly if supplied with extra nitrogen and phosphorus. As nitrogen and phosphorus are precisely the nutrients present in highest concentration in sewage outflow, several research teams are currently investigating the

possibility of carefully enriching marshes while simultaneously using them as part of a natural sewage treatment system.

Such manipulations can have unexpected consequences, however. Major nutrient inputs can change the dominant species of primary producers. Added nutrients can also increase turbidity in creeks and bays, thus lowering the photosynthetic rates of sea grasses and their epiphytes. In extreme cases, too much decomposable organic material can cause such an explosion of bacterial growth that oxygen levels in the water column drop to dangerously low levels. No major change in any ecosystem as complex as a marsh should ever be undertaken until carefully controlled experimental manipulations have been performed and their results scrutinized. A great deal of study is needed in this area, both on marshes enriched naturally—by seabird colonies, for example—and on those enriched by carefully monitored human activity.

Regardless of how nutrients and energy flow through the marsh on their own, however, many animals from nearby communities do not wait for food from the marshes to come to them.

The harsh physical conditions and the youth of temperate marshes on a geologic time scale doubly restrict the number of species adapted to marsh life. As physically stressed environments where seasonal changes in temperature and salinity can wipe out entire populations, marshes house organisms with very broad ecological requirements and with the ability to reproduce rapidly in order to invade uninhabited territory. If these systems had been around for a few million years longer, more organisms might have had the opportunity to adapt to tidal marsh conditions, and those that did so would have had the opportunity to become more ecologically specialized; as it stands, relatively few have evolved to occupy the tidal marsh.

These factors, together with high potential productivity, have resulted in marshes that combine an abundance of relatively few, generalized food species with a scarcity of large, resident predators specialized to feed on them. This fortuitous situation encourages exploitation by any organism sufficiently large and mobile to move in and out of the marsh as conditions permit. Migratory waterfowl that stop by, feed, and move on are prime examples. The whole pattern of fish life in the marshes is characterized by seasonal peaks and dips; each species moves in for a specific period of its life cycle, feeds in the marsh, and moves on.

Some fishes, like smelts and striped bass, move through estuaries to spawn in fresh water. Their larvae drift downstream to feed and grow in the estuary. While smelts often remain close by their spawning ground or travel limited distances along the coast, striped bass ultimately leave the estuary entirely to live offshore as adults.

A few species, such as winter flounder, spawn right in the open waters of the marsh; the flounder larvae and young remain there for at least two years, feeding first on zooplankton and small worms, and later on the "necks" of clams and mussels snipped off where they stick out through the mud. Young winter flounder grow rapidly and either stay inshore or move out to sea, where they frequent shallow banks and shoals before returning to estuarine spawning grounds.

Still more fishes, including most flounders, menhaden, and several members of the cod family, live in deep water as adults, in many cases far from shore. But though these species spawn at sea, their larvae instinctively head for marshes and other estuaries, where they metamorphose into juveniles and feed for lengths of time varying from months to years. What leads these tiny creatures to find estuaries and to stay in them for a finite period before leaving once more is still not known.

Even fishes that rarely enter estuaries depend

Gulls and other seabirds feed regularly on fish and shellfish that thrive in estuaries.

for food on the young of those that do; bluefish, for example, follow migrating menhaden as they cruise in and out of fertile marsh waters. Barracuda lurk around tidal inlets, snapping up any juveniles of various species that stray from the protection of the brackish water.

The habits of all of these species underscore the importance of marshes (and the estuarine bays they embrace) to animals that rarely if ever spend time in them as adults. In the wandering course of evolution, these species—many of which are important to human beings either commercially or as game fish—have come to depend on the rich, protected waters of the marsh. Although these animals forsake the shallow and turbid estuaries for more spacious open water as they mature, the marshes remain critically important to them as sources of food, shelter, and reproductive habitat during at least some part of their life cycle.

Returning to the concept of the biosphere as a superorganism, one might say that marshes serve other marine systems both as reproductive organs and as nurseries, providing nourishment and protection to organisms whose adult lives are spent elsewhere. Eliminate the marsh, or disturb it in the wrong way, and fish that depend on it to spawn will be as unable to reproduce effectively as if their sexual organs had been removed. At first, the visible effect on nearby populations may be small or even unnoticeable, depending on how long a growth cycle each species has before maturing. Time passes, sometimes even several years, encouraging a perception that the marshes are not really so important after all. Then, when the first missing year's spawn would have reached adulthood, fisheries located many miles away mysteriously begin to collapse.

Because of marshes' critical role as nurseries to so many wide-ranging marine species, isolated bits and pieces of marsh are not sufficient to maintain healthy fish populations. Although scattered wildlife sanctuaries are beautiful and restful places for naturalists, as well as being essential environments for local marine organisms, individually they are of limited value to estuarine fishes. To keep these species thriving up and down the coast, sizable healthy marshes must be allowed to remain at regular intervals along the shore. In areas experiencing rapid population growth, therefore, immediate action is needed to preserve these vital habitats.

Yet the very salt marshes whose riches attracted human populations to the coast are under intense pressure from modern society. Coastal towns have a history of casual sewage and garbage dumping. Developers eye marshes eagerly as sites on which to build yet another generation of seaside condominiums. Unenlightened visitors and even residents, encountering only the marsh's noxious products—mosquitoes, green-head flies, and the like—avidly support development proposals that nibble away at remaining marshlands.

In the United States and in Europe, marshes have been disappearing for centuries. This attrition in nursery and feeding grounds exacerbates the already serious threat to marine stocks posed by overfishing. Action to preserve the remaining marshes is desperately needed.

Ascending and descending with the tide, a mangrove snail grazes on algae coating a mangrove root.

Their air tubes conducting oxygen to sections buried in anaerobic mud, the prop roots of a young red mangrove spread to match its crown of branches.

Mangrove prop roots provide hospitable anchorages for barnacles and a host of other filter feeders.

Tropical Flooded Forests

In areas far from killing frost, salt marshes share estuaries with a curious, salt-tolerant community of shrubs and trees collectively called mangroves. First occurring on the coast of the United States as scattered clumps along with *Spartina* grasses in southern marshes, the mangrove community gradually takes over in the subtropics and tropics, until it replaces the marsh-grass community altogether. Fringing the shores of tropical islands from the Bahamas to Indonesia and straddling the deltas of great tropical rivers such as the Mekong, the Amazon, the Congo, and the Ganges, mangroves live, like marsh grasses, in the region between high tide and low tide, and they grow most profusely where rivers or streams meet the sea. Ranging in size from stunted shrubs to massive trees, the forty or so mangrove species worldwide create breathtakingly beautiful estuarine forests whose uppermost branches may tower up to 30 meters above the flooded ground in which they grow.

Gliding along winding tidal creeks that penetrate the otherwise impassable thicket of arching roots and criss-crossing branches that compose a mangrove swamp, a casual visitor would hardly equate the environment with that of a salt marsh. Instead of low-lying grasses and views of sea or land in the distance, nearly vertical walls of emerald foliage embrace the waterways on either side. In great delta forests, mangrove branches may arch together overhead, forming living tunnels that surround and protect placid creeks whose surface remains as smooth as glass even when wind-driven waves pound the coastline. Deep in the forest, the salt air seems far away, and the warm, humid air amongst the trees seems tamed and stilled by the forest canopy.

Two types of mangrove that thrive close to the water's edge—the red mangrove, *Rhizophora,* and the black mangrove, *Avicennia*—dominate the view from both creek and ocean sides. Beneath clusters of sturdy deep-green leaves are the mangroves' remarkable root systems, evolved in response to two major difficulties plants face in living along the fringes of these flooded forests. Mangroves must anchor themselves firmly against the power of wind and waves; simultaneously, they must cope with stagnant sediments in which the roots of other plants would wither and die from lack of oxygen. Faced with similar problems, but starting with different genetic potential, red and black mangroves have each evolved their own, slightly different solutions.

Red mangroves, which are exposed both to the open ocean's wind and waves and to tidal creeks' often substantial currents, anchor and buttress themselves with long, curving prop roots that grow from above-water branches and arch gracefully through the air before dropping vertically into the sediment. Aerial portions of these roots are peppered with waterproof pores that connect the open air to long, hollow passages within the root itself. These internal passages conduct essential oxygen to below-ground roots.

While anchoring plants in the sediment, prop roots also break the force of waves and currents and simultaneously provide multidirectional support for wind-tossed branches. Through some mechanism still not understood, individual *Rhizophora* plants grow root systems with different branching patterns tailored to the stresses each encounters during growth. Trees growing along the forest's ocean fringe produce recumbent branches, supported by numerous roots, to resist both winds and strong waves. Trees growing farther inland, in spots protected by other trees, grow taller, more nearly vertical trunks with far fewer roots.

Because arching *Rhizophora* roots seem to reach out from the edge of the thicket, and because—like the more northern *Spartina* grasses—they obviously stabilize and collect sediments, it was once believed that mangroves were land

Following pages:
A great blue heron surveys its domain from a perch on a red mangrove root.

builders that actively created islands. More recent work has shown that this is not the case; where additional sediment is being deposited along the margin of a mangrove forest, *Rhizophora* is likely to attempt to occupy the area of new sedimentation and in the process will ultimately stabilize it, but the preexisting forest does not cause the additional sedimentation in the first place.

Instead, mangroves respond to changes in sea level by advancing or retreating, as the tide line rises and falls. More dramatically, mangrove thickets protect the coastal zone from catastrophic ravaging by hurricanes and typhoons. During severe storms, even hardy red mangroves may be seriously damaged, but the skeletal thickets and root systems, along with marsh plants growing among the prop roots, help hold the sediments in place. Between major storms, regrowth of old plants and resettlement by new ones enable the trees to reclaim lost ground and to provide a new line of natural defense against the next meteorological assault. Red mangrove thickets thus act as dynamic coastal buffers, shrinking after major storms and expanding to their former boundaries during quiet years.

Black mangroves, which usually grow in the shelter behind an outer band of *Rhizophora,* face similar problems with oxygen-deficient sediments but fewer difficulties with the rigors of winds and waves. Each *Avicennia* plant sends out horizontal roots that run for sizable distances away from the tree just beneath the sediment surface. At regular intervals, these horizontal roots send up vertical structures called pneumatophores, whose tips grow just to the level of high tide. With upper ends perforated by pores and with interiors riddled with air channels, pneumatophores act as breathing tubes for the tree's wide-ranging, submerged root system.

Red and black mangroves' extraordinary root systems, keys to their success in a difficult situation, are also the Achilles heel that makes mangroves extremely vulnerable to unusual disturbances in their habitats. A prolonged increase in the amount of suspended sediment, a long-lasting increase in water level, or an influx of oil from an offshore spill can interfere with the pneumatophores' function sufficiently to damage or kill first the roots and then the entire aquatic forest.

Adding to the mangrove swamp's personality are the scores of estuarine animals that use mangrove thickets, both above and below water, for home, shelter, feeding, or resting places. Birds by the score, including permanent residents and seasonal migrants, call and fly among the branches or hunt and peck through the mud around the roots, disappearing into the thicket at the slightest disturbance. Hundreds of crabs scamper up, down, and across arching roots and trunks just above the water, doing their squirrel-like best to keep a branch between their bodies and any observer. Low tide exposes the creek banks, their mixture of sand and peat riddled with the burrows of tropical fiddler crabs.

Below the water line, filter-feeding mussels and oysters cling to *Rhizophora* roots that project into plankton-carrying currents the tides sweep in and out. Between oyster shells, root surfaces are carpeted with algae, bacteria, and small, sedentary invertebrates. Limpets cling to patches of clear bark, and periwinkles wend their way among larger attached animals and plants, grazing on the rapidly growing algae. Powder-fine silt caught by algal filaments and rough-surfaced oyster shells houses myriad tiny shrimp and worms. Schools containing hundreds or even thousands of inch-long minnows dart along the creeks and dive into the root thickets for protection. Mosquitoes and biting flies sometimes fill the still, evening air.

For all the obvious physical differences between mangrove swamps and salt marshes, these

Stiltlike pneumatophores reach out of the mud to provide air to submerged black mangrove roots.

tropical salt-water forests house food webs and ecological interactions remarkably similar to those of their temperate-zone counterparts. As is the case in salt marshes, relatively little plant material is lost to mangrove herbivores such as deer, mangrove tree crabs, caterpillars, and grasshoppers. Detritus plays the central role in connecting above-water plant growth with the swamp's marine inhabitants. As new leaves grow in the canopy, old ones drop into the mud among the tangled roots. There they are immediately attacked by decomposers, beginning the same sort of detrital-bacterial food chain as the one based on *Spartina* foliage. As the smell of rotten eggs in mangrove sediments clearly proclaims, sulfur-rich bacteria grow in abundance and may be critical organisms in several bacteria-based food chains.

Just as happens in marshes, many of the nutrients that support mangrove growth come from fresh-water drainage, both through rivers and streams and through percolating ground water. Like marshes, too, many mangrove areas seem to be nitrogen-limited and respond to localized inputs of added nutrients with faster growth and lusher foliage. In sparsely settled Caribbean islands, where residents of the scattered homes suspended on stilts over mangrove channels drop their wastes into tidal creeks, trees along the banks, and their associated root communities, respond to the extra nitrogen by growing noticeably larger. The same reaction is evident where dense colonies of nesting seabirds have enriched tidal creeks with their droppings. Where human habitation becomes more concentrated, however, sewage dumping supports mangrove growth only if the tidal flushing is sufficient to aerate and exchange water regularly. Enrichment (as those who study sewage addition to such systems like to call the process) in stagnant areas can lead to rampant bacterial growth and to conditions that actually inhibit mangrove growth.

Work on energy flow, internal nutrient cycling within the mangrove area, and nutrient flow into adjacent habitats is still largely in its infancy; ecologists do not understand these systems nearly as thoroughly as they do salt marshes. Given the diversity of mangrove ecosystems, it is to be expected that internal carbon and nitrogen cycles and nutrient export rates are highly variable, both from year to year in a single swamp and in the same time period among different systems. Little is known conclusively about either the extent or importance of passive nutrient transport from mangroves to adjacent sea-grass and reef systems.

Similarities in food-web structure notwithstanding, the mangrove fauna is naturally somewhat different from that of temperate-zone marshes. Mangrove shrimp, crabs, and fishes do not belong to the same genera and species as their temperate-zone counterparts. The mangroves' cast of characters is also a bit larger. This is not surprising; as a rule, tropical ecosystems contain more species than temperate ones. Compared with temperate-zone marshes, mangrove areas are often considerably older, having escaped recent glaciation, and are physically less rigorous, being situated beyond reach of killing frost. These conditions encourage the evolution of more complicated, highly evolved interrelationships among species, and over time they afford opportunities for more organisms to adapt to mangrove environments.

Rhizophora roots, for example, are widely infested with varieties of worms, crustaceans, and even molluscs that bore through wood and cause substantial damage to plant tissue. One's first reaction would naturally be to assume that this situation is strictly detrimental to the mangroves. Some researchers in Florida's Everglades who agree with this diagnosis link higher infestation levels to rising levels of salinity due to diminished fresh-water supplies in the swamp. Other work, however, shows that at least a moderate level of infestation

In extensive mangrove areas like the Florida Everglades, food webs include large, top-level carnivores like alligators.

Sea turtles spend much of their time elsewhere, but they return regularly to estuaries where they feed on submerged seagrasses.

causes growing roots to branch more frequently, a change in growth habit that is believed beneficial to individual plants. As yet, the full nature of the interaction between borers and mangroves is still unresolved; it may be a simple case of parasitism, or it may be part of a complex natural regulatory process in which interactions between trees and borers help the trees maintain a balance between growth and decay.

Despite the presence in mangrove swamps of species whose interactions are complex, however, the ecological roles many organisms play are nearly identical to those of similar species in temperate marshes. Although mangrove swamps harbor more species than marshes do, when the number of these species is compared to the thousands of species of tropical fresh-water organisms or with the bewildering variety of plants and animals living on coral reefs a short distance away, the mangrove fauna appears to be proportionately as species-poor as that of any temperate estuary.

Mangroves regularly shelter killifishes, mosquitofishes, and several relatives of aquarium livebearers, the platyfishes. Most of these are ecological generalists whose foods include detritus, mosquito larvae, algae, and small crustaceans. Like salt-marsh species, permanent residents of mangrove swamps are often present in staggering numbers. These small species in turn are eaten by larger predators—including snook, ladyfish, tarpon, gars, snappers, and such wading birds as herons, ibises, and storks—that travel in and out of the swamp to feed. A wide variety of other animals, including pink shrimp, spiny lobsters, blue crabs, mullet, menhaden, red drum, spotted sea trout, jacks, and groupers—all commercially valuable—enter mangrove lagoons periodically to feed on the abundant small prey.

Some mangrove invertebrates, including tropical fiddler crabs, are sufficiently closely related to temperate-zone species to be placed in the same genus. Other mangrove dwellers, including several detritivorous shrimp species, are not closely related to corresponding temperate-zone animals by evolutionary descent but are practically carbon copies of each other in their ecological roles. Still other animals, such as the ubiquitous mangrove tree crabs, are specific to the mangrove community. Nonetheless, mangrove crabs, too, are quite catholic in their tastes, grazing on mangrove leaves and devouring any insect larvae they come across. The crabs themselves fit into food webs in various places; their planktonic larvae are the prey of many fishes, while juveniles and adults are relished by ibises, raccoons, larger members of their own species, and—if small ones fall into the water—such large fishes as mangrove snappers.

Ecological similarities to other estuarine systems continue in mangrove lagoons and tidal channels that are open to the sun, where genera of aquatic plants such as *Thalassia* and *Zostera* fill the niche held by *Ruppia* in temperate salt marshes. Blessed with abundant sunlight and consistently high temperatures, tropical sea grasses grow rapidly. Almost as quickly as new leaf tissue is added at the base, older tissue above is overgrown with epiphytic microorganisms, sedentary invertebrates, and algae that often equal or exceed the mass of the leaves themselves.

Both sea grasses and their epiphytes often grow so quickly that they would eventually fill their lagoons were they not equally rapidly attacked by herbivores. Turtle grasses got their name from the fact that sea turtles in large numbers found their sustenance in the grass beds before humans with a taste for turtle soup decimated their populations. Like many other estuarine feeders, turtles do not actually live full-time where their food grows; they spend their nights in deep holes or in caves on the ocean side of nearby coral reefs, surfacing about once an hour to breathe. However, during the hours of twilight—both dawn and dusk—they head for sea-grass beds to feed.

Each turtle returns regularly to its own specific feeding site, where constant grazing keeps the grasses cropped to a length of only a few centimeters. Chemical analyses have shown that freshly grown tissue at leaf bases is higher in nitrogen and lower in fiber content than older leaves, so turtles maintain a crop of superior leaves by feeding in this way.

Manatees, like turtles, are submarine grass feeders that once ranged over far greater areas than they do today. Slow-moving, gentle herbivores, manatees eat up to 20 percent of their body weight in vegetation daily. Since these animals weigh up to 500 kilograms, a herd of them can consume the daily production of a sea-grass meadow with little trouble. Manatees feed differently from turtles; instead of cropping grasses near the roots with chopping bites, they use their facial bristles to dig into sediments and uproot the grasses by their long, connecting rhizomes. Shaking their heads vigorously to dislodge attached sediments, these sea cows then chew and swallow the plants whole.

In addition to supporting these highly visible grazers, plant material produced in sea-grass meadows acts as a vital, living bridge between mangrove swamps and the coral reefs that often surround them. Plant-eating reef dwellers, from parrotfish to surgeonfish and porgies, migrate regularly to the grass beds to feed on both the sea grasses and their epiphytes. The combined mangrove swamp and sea-grass meadow habitats provide marine life with important nurseries and spawning grounds, much as temperate marshes do; scores of fishes and invertebrates migrate in and out seasonally to breed and to grow as juveniles.

Panulirus, the spiny lobster whose flavorful tails delight gourmets the world over, rely totally on mangrove swamps as habitats during critical parts of their life cycles, although adults are rarely if ever found there. *Panulirus* subadults prefer to spend their daylight hours hiding in nooks and crannies on shallow coral reefs, wandering over into adjacent grass beds to feed at night. Sexually mature females mate and then migrate to deep offshore reefs, where they stay until their tiny, planktonic larvae are ready for release. *Panulirus* larvae, which look nothing like their parents, enter the plankton and drift, mostly at the mercy of the currents, for between three months and a year, when a final larval molt transforms them into tiny, transparent, miniature adults.

At this point their movements suddenly become directed, though how and by what mechanism no one knows. By night they migrate toward mangrove swamps along the coast; by day they burrow in the bottom. Reaching the mangrove thicket, they settle on algal mats in tidal creeks, on mangrove roots themselves, or on floating mats of algae within the swamp. After another molt and the acquisition of body pigments like those of their parents, the juveniles become bottom dwellers, moving into grass beds where they feed and grow for as many as nine to twelve months before traveling to inshore reefs to begin the cycle again.

The commercially vital pink shrimp, *Penaeus,* is just as tightly linked to mangrove–sea-grass systems. While they are not actually caught in estuaries, these tasty invertebrates spend their entire juvenile lives moving back and forth between mangroves and sea-grass beds in bays and rivers. The great shrimp fishery of the Tortugas Banks, for example, is nourished through infancy in the vast reaches of the Florida Everglades. Only after reaching maturity do these animals move offshore, where they are harvested by commercial fishing fleets.

Humans impose arbitrary divisions on various habitats; these have no meaning for fishes. Fishes wander back and forth from one system to another, according to their needs. Gray snapper juveniles,

Grazing mammals that abandoned land for sea grass pastures long ago, manatees consume prodigious quantities of submerged vegetation (© Douglas Faulkner).

as an example, first live in sea-grass beds and later move up into mangrove thickets. As adults, they move once more, heading off into deeper water where they frequent coral reefs. Numerous so-called fresh- and salt-water species do not in fact frequent one habitat exclusively but move to feed wherever conditions suit them.

In places where mangrove forests are fed by large rivers, heavy rains during wet seasons allow fresh-water fishes such as gars, sunfish, bass, and catfishes to descend into the swamps to feed. During dry seasons, on the other hand, increasing levels of salinity encourage more typically marine species such as jewfish, stingrays, needlefishes, jacks, and barracuda to ride the incoming tides up into lagoons and creeks in search of prey.

In places where mangroves enclose large bays and lagoons, and strong tidal flushing keeps salinities near that of the sea, large sea-grass meadows encourage regular visits by still more marine species, including drums, porgies, mojarras, mullets, flounders, sole, anchovies, and herrings. Here, visiting reef fishes, including surgeonfish, doctorfish, goatfish, and parrotfish, also use mangroves and grass beds as nurseries. Arrivals of juveniles and larvae of such species in Florida peak in spring and early summer, and the animals reach their highest population densities in late summer and early fall. In late autumn and winter, subadults move off into deeper water as temperatures in the shallow swamps begin to drop.

Aquatic animals are not the only ones to shelter seasonally in mangrove communities. Studies of swamps in Jamaica and Mexico have shown that many familiar North American songbirds winter there. The classification of these peripatetic species, of course, is another example of the human tendency to categorize on the basis of a provincial outlook; although New Englanders may view warblers and the like as native northern birds that merely winter "down south," Jamaicans just as naturally view them as indigenous species of their winter fauna that just happen to travel north to breed. The message that comes through time and time again is that living organisms, nearby habitats, and even far-off ecosystems are connected with one another in ways that have taken eons to evolve. The biosphere as superorganism is a most useful concept because it reminds us, among other things, that local human activity of any sort that affects one environment may have unforeseen consequences elsewhere.

Jacks often ride incoming tides to prey on estuarine fishes and invertebrates.

Producers and Consumers

Kelp: The Fluid Forest

Flamboyant Garibaldis jealously guard feeding territories within and around kelp forests.

The emerald-green sea off the California coast is cold yet laden with life. From its rocky bottom, giant brown algae called kelps stretch ribbonlike toward the surface, creating a shadowy, fluid forest of arching fronds whose tips wash silently back and forth with the waves. Beams of sunlight penetrate the undulating canopy, spotlighting passing schools of fishes that pause inquisitively before detecting movement nearby and darting into the protection of the algal thicket. Across the bottom, crabs prowl and urchins graze through clumps of smaller kelp species and over scattered rock piles.

Kelp forests evoke from marine biologists the same admixture of awe and admiration, the same tendency toward superlatives, that giant redwood forests evoke from terrestrial biologists. Giant kelps such as *Macrocystis* are arguably the world's largest algae, stretching 20 to 40 meters between bottom and surface, and often trailing in the waves for some distance thereafter. Kelps also grow with phenomenal rapidity; a mature *Macrocystis* frond can add up to 50 centimeters of new tissue on a good day, leaving even the most energetic terrestrial plants far behind. Kelps' sheer physical presence creates a hauntingly beautiful, three-dimensionally complex environment that provides both food and shelter to a thriving and diverse community.

In contrast to tropical organisms, which are confined to equatorial regions by intolerance to cold, kelps are strictly cold-water species that rarely survive in locales where water temperatures rise above 20°C. Excluded from warm water, kelps grow most luxuriantly in places such as the Pacific coast of North America, where the Alaska and California currents veer offshore into the Pacific, creating an upwelling of chilled, nutrient-laden bottom water that stretches from the Aleutian Islands off Alaska around and down into southern California. Other kelp forests flourish under similar conditions off the coasts of Japan and Siberia, off the Atlantic coasts of Canada and New England north of Cape Cod, off the west coasts of South America and South Africa, and off the northern coasts of Great Britain and Scandinavia. Spread throughout the world's northern and southern waters in this way, kelp forests neatly complement the distribution of coral reefs, which survive only where water temperatures stay reliably *above* 20°C.

Throughout the kelp zones of the Pacific and the South Atlantic, *Macrocystis* dominates the near-shore landscape. Its tall, narrow fronds grow in tight clusters but space themselves out to create open, roomy forests through which aquatic animals swim with ease. On both sides of the North Atlantic, however, *Laminaria* predominates, its closely packed growth forming dense underwater thickets more amenable to crawling than swimming.

No matter where they grow, kelp forests are a study in intricate interaction between algal communities and the animals that inhabit them. Even the mightiest kelp plants, for example, begin their lives as microscopic plantlets that germinate from spores lucky enough to land on a rocky substrate in a spot where enough light penetrates to satisfy their energy requirements. At this stage, any algae eaters passing by can gobble the plants up in their entirety, a fact that makes the density of herbivores in an area a matter of primary importance in determining whether or not new plants can establish themselves.

From these humble beginnings, kelps grow rapidly in two directions simultaneously. A flattened, leaflike structure pushes upward in search of light, while a spreading, flattened plate of cells spreads out and clasps the bottom. The infant leaf turns into a long, flexible stipe that stretches eagerly toward the sun, bearing with it the flattened blades that carry on photosynthesis. Spreading out at the surface, where they receive as much sunlight as possible, kelps form a canopy much like that of a terrestrial forest. Below, the anchoring cell mass differentiates into ropelike haptera that look similar to roots of higher plants. Haptera have no nutrient-procuring function, however; they serve exclusively as holdfasts to attach the algal youngster firmly to rocks underneath.

As plants grow, their appearance changes further, with different species taking on different growth forms. *Macrocystis* fronds produce a gas-filled float at the base of each blade to buoy up its photosynthetic surfaces. The entire frond—stipe, floats, and up to 200 blades—grows as a unit, and lasts about six months. Replacement fronds grow in pairs, rising toward the surface at the astonishing rate of 3 to 5 meters each week under good conditions. Plants are perennial, usually living for between three and seven years before they die from any of a variety of causes. *Laminaria* fronds, on the other hand, grow in the manner of conveyor belts, with new tissue appearing at the base as older tissue at the leaf tip becomes senescent and erodes away.

Haptera, too, are produced as the plants grow, continually expanding and increasing the plants' points of attachment. In mature *Macrocystis* plants, clumps of fronds sprout from a massive holdfast network composed of a dead, dark brown area in the center, a middle ring of woody-looking holdfasts still clinging to the substrate, and an outer layer of yellowish brown, actively growing haptera on the outside.

Over, under, around, and through the kelp forest live myriad organisms that use kelps for food, shelter, or both. The dense frond network tames currents and drains the force from all but the worst storm waves, providing convenient shelter for both fishes and marine mammals such as otters and seals. The presence of these treelike algae creates an intricate, three-dimensional, living framework that enormously increases the volume of useful habitat in the area. Fish species ranging from blennies, pipefish, and eels to the striking Garibaldi, as well as a wide range of large fishes normally associated with rocky areas in much deeper water, find the kelp labyrinth a perfect environment. Without the kelp forest for a habitat, many of these animals either would be absent or would be present in much smaller numbers.

While constantly growing, kelp plants are also constantly eroding, producing a steady stream of detritus that plays nearly as significant a role in the kelp forest food web as it does in the food webs of salt marshes and mangrove swamps. Inside kelp groves, in fact, live a species of filter-feeding mussels—one of the few multicellular organisms that produce enzymes capable of digesting cellulose. This species is thus able to feed not only on the

Masses of rootlike haptera anchor giant kelp to rocks on the ocean floor.

Right:
Stretching to over 40 meters in length, giant kelp create roomy undersea forests off the Pacific coast of North America.

Gliding along kelp stems, moon snails (*above*) and sea urchins (*below*) feed on the outer layers of plant tissue.

Poised on a blade of decaying *Laminaria,* an Atlantic sea star glides along on its hundreds of tube feet.

Following pages:
Floating lazily among kelp blades, a California sea otter makes a leisurely meal of a sea urchin (© Jeff Foott).

Kelp blades extract nutrients from the surrounding water as they photosynthesize. The holes and ragged edges bear witness to the depredation of herbivores.

Common in both Atlantic and Pacific oceans, temperate nudibranch species dine on hydroids that overgrow the kelp blades and stalks.

bacteria that grow on kelp fragments, as other detrital feeders do, but on the kelp particles as well.

Other filter-feeding organisms attach themselves to kelp blades and haptera, at times growing over the new blades almost as quickly as they are produced. Sponges spread through mats of intertwined haptera, which also shelter scores of tiny shrimp, crabs, lobsters, brittle stars, and small fishes. Abalone are common in Pacific-coast kelp habitats, while American lobsters, *Homarus americanus,* are denizens of North Atlantic kelp beds.

Though most organisms that live on kelps are epiphytic filter feeders, these giant algae are far more succulent than either *Spartina* grasses or mangroves, and they provide food directly to numerous grazers. Ponderous kelp snails crawl up stipes, scraping off the outer layer of living cells as they go. Burrowing through the matted haptera, hordes of termitelike *Phycolimnoria,* commonly called gribbles, gnaw their way through the kelps' vital holdfasts. Gribble action often weakens holdfasts so seriously that storm waves can uproot the plant and carry the whole assemblage away. In stable kelp forests, in fact, gribbles may be the major source of mortality among mature kelp plants.

Human hunters and fishers, unfortunately, have substantially upset the fragile ecological balance in many kelp forests off both the Atlantic and Pacific coasts of the Americas. The first casualty, the slow, gentle Stellar's sea cow, which once grazed on the lush kelp growth of the Pacific Northwest, was hunted to extinction in the eighteenth century. Almost succumbing to the hunt was the Pacific sea otter—once a common sight on the west coast, and now reappearing under protection from fur traders. On the Atlantic coast, American lobsters are so heavily fished today that their near-shore population is but a fraction of its former size and includes almost no large individuals.

Interference of this sort has caused unexpected problems for kelp ecosystems. Otters and lobsters are important carnivores in their respective habitats; sea otters relish abalone and sea urchins, while lobsters devour almost anything they can get their claws on. As is often the case in complex ecosystems, the predatory effects of these top-level carnivores are of critical importance in controlling the populations of the animals on which they feed. Sea otters and lobsters are prime examples of keystone predators, so called because their presence at the top of the food web maintains ecosystem stability and their removal brings the system tumbling down. At the core of these predators' stabilizing influence in kelp ecosystems is their predilection for the ubiquitous sea urchins of both coasts.

Sea urchins are remarkably efficient grazers, and kelp is one of their favorite foods. While wave action and the activity of predatory fishes usually keep sea urchins from doing excessive damage to the upper parts of the kelp stipes and canopy, few predators stand between these eating machines and the plants' critically important holdfasts. As sea otter and lobster populations declined, the sea urchin population exploded. Wave after wave of the spiny herbivores moved through the kelp forests, nibbling on haptera and weakening the kelps' anchorage until storms uprooted the plants and washed them ashore.

Combined with naturally occurring perturbations like disease and climatic changes, these major alterations in grazing pressure may have been prime movers in the decline of kelp forests over the last half century.

On the west coast, few kelps are left in some areas where acres of them once nourished abundant sea cow populations. Off the coast of Nova Scotia, as the lobster population dropped 50 percent in the 1970s because of overfishing, an estimated 70 percent of the kelp beds disappeared. High densities of sea urchins remain in denuded areas, making it nearly impossible for the micro-

Following pages:
Common bat stars crawl among kelp holdfasts and colonies of red algae in a California kelp bed.

Cowries, like their shell-less cousins, the nudibranchs, often include in their diets such organisms as sponges, which contain toxic chemicals that deter other would-be predators.

Right:
A gas-filled bladder at the base of each *Macrocystis* blade holds the long, trailing stalks in sunlit water near the surface.

scopic stage of the kelp life cycle to become established before the plants are devoured. Although the sea urchins wreak havoc around them, they usually survive; most are generalist feeders, and they simply switch to smaller algae when kelps are no longer available. Members of the kelp fauna that are specifically dependent on kelps for food or shelter, however, are not so fortunate.

The future of kelp forests on both coasts is still uncertain. Where sea otter populations have recovered—around Amchitka Island in the Aleutians, for example—their diligent hunting has reduced sea urchin populations and allowed kelps to reestablish themselves and grow densely once again. In certain locations off the California coast, applications of quicklime have at least temporarily lowered sea urchin populations and made some kelp resettlement possible. In the Atlantic, a recent epidemic among sea urchins caused mass mortality among these grazers; here and there, kelp beds have started the long, slow process of regrowth.

In the meantime, the lobster fishing situation in former kelp bed areas has become a catch-22. Lobsters, which helped maintain the stability of the ecosystem in which they thrived in large numbers, have been grossly overfished. As lobster populations declined, the ecosystem that sheltered and nourished them in high densities has nearly disappeared in some places. Without the kelp habitats and the food species that flourish in them, lobster populations remain low, causing fishermen to redouble their efforts to keep the restaurant tables laden. But the more heavily lobsters are fished, the less effective they are as a control of urchin populations, and the more difficult things become for kelps. Whether these systems can establish a viable new equilibrium under conditions that include human predators or whether human beings must simply change their ways to ensure the survival of kelps is a question that can only be answered over time.

Carnivorous starfish often feed on several trophic levels; their activity often keeps the food web in balance.

Plankton: Life Adrift

Defying the shifting, unpredictable wind with aerial grace, gray and white shearwaters skim deftly over the North Atlantic, their wingtips coming so close to the waves that they seem to cut right through the whitecaps. Around them petrels fly, relying on keen eyesight and quick reflexes to snatch small prey from just beneath the surface. Between birds and horizon, plumes of spray and water vapor mark the measured breaths of finback, humpback, and minke whales cruising the banks in search of the tiny fish they must devour by the ton in order to survive. These animals are not alone in their pursuits: the turbid, blue-green sea is dotted with fishing boats ranging from sleek craft for chartered sport-fishing to massive commercial trawlers.

So far from shore, no familiar attached plants can serve as primary producers for the food web that supports all of this life; the surfaces to which plants might attach themselves are deeply submerged and are shrouded in darkness by the light-absorbing properties of the overlying water. Here, therefore, all higher life depends ultimately on the diverse assemblage of water-borne plants and animals collectively called plankton or "drifters" (because most of them float at the mercy of the currents). Many plankton are single-celled and microscopic, and many others are just large enough to see with the naked eye; still others may be half a meter across. Under the microscope or hand lens, some are recognizable as tiny shrimplike creatures with delicate, transparent shells. Some have the simple structure of diminutive basketballs; others resemble ornately carved pillboxes; and others produce extravagantly decorated shells that look for all the world like lunar landing capsules from an old Buck Rogers movie.

The floating world of the plankton is alien, challenging, and difficult to comprehend for creatures as determinedly terrestrial as ourselves. This dynamic, multilayered, floating ecosystem stretches across the entire surface of the sea, and down into the depths for hundreds of meters. Open-water ecosystems include thousands of plant and animal species, each with its own light, temperature, and nutrient requirements, its own growth characteristics, and its own relationships with nutrients and predators. Categorizing such a system and its inhabitants from the confines of a surface-bound research vessel poses formidable problems, and oceanographers' knowledge about open-water plankton continues to grow and change rapidly.

The open waters of tropical seas offer plankton regular doses of intense sunlight but often carry nutrients in quantities too low to be measured by analytical techniques. Lean phytoplankton growth in such regions apparently supports even sparser growth of the small, shrimplike zooplankton that graze on phytoplankton. Starved for nitrogen and phosphorus, the crystalline blue waters of the mid-oceanic photic zone thus look superficially like the oceanic equivalent of a desert—an area inhabited only by scattered phytoplankton whose growth is so nutrient-limited that they subsist only by growing at the slowest rate possible to support life.

Only tiny primary producers, called nanophytoplankton (from the Greek *nanos,* meaning "dwarf"), which grow little larger than 20 microns in length, can thrive in these nutrient-poor waters. So small are these primary producers that they slip through the filtering apparatuses of most large zooplankton. Such tiny cells can be eaten efficiently only by microzooplankton—very small zooplankton, 20 to 200 microns in size. At the next trophic level, these animals fall prey to macrozooplankton, 200 to 2,000 microns in size, which in turn provide meals for megazooplankton such as open-water shrimp, measuring over 2,000 microns in size. Shrimp are a primary source of food for small, open-water fishes, and these fishes subsequently fall prey to salmon, tuna, and squid.

Therefore, when it comes to producing the kinds of large animals that interest people as food, open oceans must labor under a double handicap. In the first place, they have few nutrients available to support primary production, which places a low ceiling on the total amount of living tissue the area can carry. On top of that, a food web involving several energy-wasting steps between primary producers and final consumers ensures that the final trophic level can contain relatively little living tissue, which translates into few large predators.

However, several planktonic ecosystems in various parts of the world are dramatically more productive than those of the open sea. In areas along the coasts of major continents, fresh-water runoff from the land provides enough nutrients to allow larger primary producers, the microphytoplankton, to join the smaller forms. In these richer pastures, several phytoplankton groups flourish in sufficiently large numbers to color the water green with their chlorophyll.

Among the most abundant and interesting of these plankton are the dinoflagellates: single-celled, photosynthesizing protozoa whose pair of whiplike flagella enable them to swim around with great speed. Because many dinoflagellates can form dormant cysts when conditions become unfavorable and then break out of those cysts to reproduce with astonishing speed when good conditions return, they respond to good growing conditions by producing sudden blooms of phytoplankton.

Mixed in with the dinoflagellates, diatoms flourish in large numbers. Individual and pillbox-shaped or as long chain formers with elongated spines, free-floating in the water or attached to the bottom or to larger algae, diatoms are large enough and numerous enough in the coastal zone to support a food chain noticeably different from that found in the open sea.

Because the mixture of nanophytoplankton and microphytoplankton found in coastal waters can be either eaten directly by microzooplankton or filtered from the water by clams, mussels, worms, and other bottom-dwelling filter feeders, food chains in these areas are often two or more steps shorter than those of open water. Microzooplankton are the preferred food of such sizable free-swimming fishes as alewives and menhaden, while filter feeders are eagerly seized by cod, haddock, and other large, bottom-feeding fishes. In both cases, only three trophic levels span the gap between primary producers and consumers of reasonable size. Even larger animals such as bluefish, sharks, and people are then only four trophic steps from the source. Both higher primary productivity and a shorter food chain work together here to support more higher-level consumers.

Still more productive planktonic systems thrive above shallow banks in the storm-tossed North Atlantic, in the Pacific off the coast of Peru, and in other sites where winds, bottom topography, and ocean currents interact to bring nutrients from the ocean floor up into the photic zone. The turbid, vividly green water of these upwelling zones—a far cry from the turquoise of the open sea—attests to the concentration of primary producers that flourish here.

A continuous supply of nutrients nourishes yet another variety of phytoplankton community, one dominated by the largest chain-forming diatoms. Linked together in strands several cells long, these chain formers are large enough to be eaten directly, not only by the largest macrozooplankton—the shrimplike krill of the southern oceans, for example—but by small fishes such as anchovies, as well. Krill and anchovies, of course, are large enough to be worthwhile prey for everything from seabirds, seals, and whales to human beings. Upwelling areas thus combine high productivity with food chains between two and three

Shallow-water amphipod shrimp such as these rarely grow larger than a centimeter.

This open-water amphipod, collected in deep water, begins life as small as its shallow-water relatives but may grow up to 15 centimeters long (© L. P. Madin).

Far more important than once believed, gelatinous zooplankters float near the surface in huge swarms.

Left:
Most crustaceans, even large ones, spend part of their lives in the plankton; this larva from the open sea belongs to a relative of the familiar lobsters and shrimp (© L. P. Madin).

trophic levels long to support the greatest amount of large animal tissue of any planktonic system.

Biologists estimate that in years when upwelling currents are strong off Peru's coast, up to a seventh of the anchovies caught worldwide by both people and seabirds come from this area, a mere two hundredths of 1 percent of the world's oceanic area. Several small upwellings along the Pacific coast of the United States provide locally productive, though more limited, fisheries there, and island nations such as Japan have for centuries harvested the riches produced by their upwellings.

It is thus no accident that people, whales, and seabirds congregate on George's Bank and places like it despite cold, turbid water and often frigid winter temperatures. Starting with primary productivity about six times higher than that of the open sea, these areas end up yielding large animal tissue at a rate in excess of 36,000 times that of the open sea. For people interested in harvesting the oceans, therefore, these limited areas that so dramatically out-produce the rest of the world ocean are invaluable commercial zones deserving both husbandry and legal protection.

The ecological importance of planktonic systems, however, goes far beyond their role in supporting human beings and other large creatures. To list the animals found at one time or another in the zooplankton would take several books. It is a curious fact of life in the sea that an overwhelming number of animals spend at least part of their lives in the plankton. Corals, barnacles, and mussels, which in their adult forms are literally glued in one place, find in planktonic larval forms the only way to disperse their genes and colonize new territory. At spawning time, these animals release hundreds or even thousands of tiny eggs and sperm cells, which unite with each other in the water column and scatter in the currents shortly thereafter. Virtually all crabs, shrimps, and lobsters have larval stages that feed and grow as part of a drifting planktonic cloud before settling down to lead their adult lives. Even fishes that are free-swimming as adults, such as herring and eels, have tiny, planktonic larvae (and almost as small early juvenile stages) that join the rest of the planktonic crew for the first few weeks or even months of their lives.

The health of these free-wheeling drifters is thus critically important, not because of the role they play directly as objects of commercial fisheries, but because any disturbance to their community can cause unexpected changes in the populations of animals whose adults are anything but planktonic.

Recognizing the far-reaching importance of plankton, scientists are proceeding with research into these little-known oceanic ecosystems with all due speed. Plankton biologists today combine field work with insights, conceptual advances, and chemical techniques developed over many years in the laboratory. Some modern researchers even forsake the shelter of their research vessels to enter the realm of oceanic plankton with diving gear to collect specimens by hand.

The effort required to enter the world of individual plankters has been more than repaid in new and exciting insights into the dynamics of open-water life. Several important components of the plankton slip right through plankton nets, while others—gelatinous zooplankton called ctenophores—are transparent will-o'-the-wisps so fragile that net collecting damages them beyond recognition.

Scuba divers, however, can spot individual, intact ctenophores by the way they reflect and refract light in the water as their rows of waving cilia break sunbeams into shimmering rainbows. These animals would never survive even the most gentle of standard collecting techniques, but a careful diver can gently herd them into jars. Even such single-celled forms of zooplankton as foraminiferans can be collected in a similar fashion, as can clusters of floating algal filaments and marine snow (collections of translucent, drifting organic particles).

New studies of these previously little-known gelatinous zooplankton have turned up hitherto unsuspected ecological threads in oceanic food webs. Many gelatinous zooplankton are adaptable organisms that grow and reproduce extraordinarily rapidly when conditions permit. They feed by secreting mucus filters that can trap nearly anything from large diatoms to the smallest bacteria. Some produce and maintain a mucous net inside their mouths and use it to filter water pumped through their bodies; then they swallow the net, particles and all. Others, called larvaceans, build spherical houses of mucus many times the size of their bodies, through which they pump water by wagging their tails. These outside filters concentrate particles that are removed and periodically swallowed. Larvaceans do not eat their mucous houses; when a house becomes too clogged with debris to function, the animal casts it off and secretes a new one.

Both of these feeding techniques allow gelatinous zooplankton to filter enormous volumes of water very efficiently, a necessity in the sparsely populated open water where they live. Because their bodies are at least 95 percent water, some gelatinous zooplankton, such as salps, can grow in size and reproduce very rapidly, producing swarms that can stretch over miles and contain up to 500 individuals in each cubic meter of water. Fitting into the intricate web of life in the sea in numerous ways, salps eat bacteria too small for most sizable plankton to capture and make that source organic matter available to their own predators. On the other hand, salps produce fecal pellets from planktonic food in abundance, increasing the flow of nutrients down and out of the photic zone. And, perhaps most interestingly, the discarded mucous houses of larvaceans, often containing up to 50,000 trapped, living phytoplankton cells, immediately become homes for bacteria and end up as marine snow particles.

Scuba collections have also yielded intriguing revelations about foraminiferans, the delicately constructed, single-celled zooplankton whose fragile bodies were damaged beyond repair in collecting nets and which had never been studied alive. Forams, often summarily treated in a sentence or two even in marine ecology texts, exhibit an impressive array of evolutionary adaptations to their floating lifestyle. Their spherical, calcium carbonate shells are usually no larger than a millimeter in diameter, but many species are covered with a network of radiating spines that increases their diameters up to ten times. Along these spiny supports, foraminiferans spread dense, sticky nets of living cytoplasm called rhizopodia. Since a spherical organism's ability to trap suspended particles increases as the square of its radius, spines and rhizopodia together increase forams' food-catching ability hundreds of times.

Even the way forams feed is unexpected. Far from being exclusively passive interceptors of detritus, bacteria, and algae, many tropical forams actively prey on shrimplike copepods nearly their own size. Rhizopodia act first as a spider's web, snaring prey and dragging it in tightly toward the shell. Then these flexible but powerful strands invade the prey's body en masse, rupturing the prey's muscle tissue into fragments, which are then engulfed and wrenched bodily out of place. Captured bits of tissue fragments are pulled into the foram's cell body where they are digested, while the empty, indigestible shell is discarded.

Many foraminiferans from nutrient-poor waters maintain ecologically efficient relationships with small dinoflagellates that live symbiotically within the forams' cytoplasm. By day, the dinoflagellates migrate out along the rhizopodial fila-

A free-swimming relative of snails and nudibranchs, this pteropod is a common member of the plankton of New England.

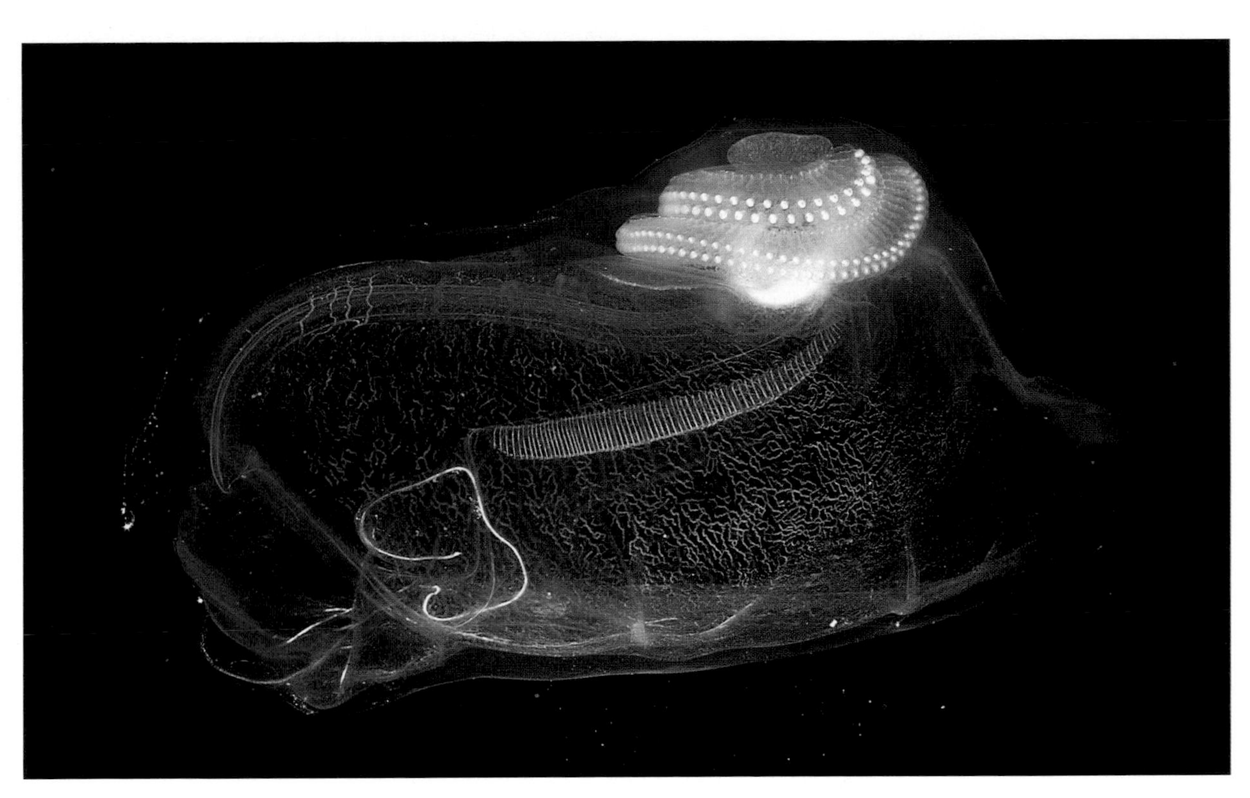

Arranged like a luminescent lace collar, a string of baby salps produced asexually by budding rides atop the parent. Inside each individual—all of which are female—is a single egg. On maturation, the egg is fertilized by a male and released, whereupon the female salp changes to a male and begins to reproduce asexually once again (© L. P. Madin).

ments, seemingly in order to maximize their exposure to sunlight. By night, they withdraw deeply into the central shell. Difficulties involved in keeping planktonic forams alive in the laboratory have not yet permitted researchers to investigate the chemical basis of this association thoroughly. Nonetheless, experiments clearly indicate that forams hosting dinoflagellates and kept in normal sunlight grow, divide, and reproduce more luxuriantly than do forams kept in the dark or those whose algae have been chemically eliminated.

Despite the absence of chemical evidence, it seems clear that some exchange of materials is going on between forams and their guests. Because they are animals, forams generate carbon dioxide and ammonia as waste products. While toxic to the animals themselves, these wastes are perfect nutrients for dinoflagellates. And because the algae grow right inside the animal cell where the wastes are generated, they are able to absorb the compounds with a minimum of effort. The nutrients do not get diluted in the surrounding sea water first, so the plants do not have to reconcentrate them.

This self-contained fertilizing and waste-removal system is extremely efficient for supporting life in nutrient-poor water; both primary producer and animal are located in the same cell. Although the full story of chemical exchange between forams and their algal symbionts is still not documented, relationships that seem to be similar show up repeatedly in tropical seas. Energy and nutrients flow between animal and plant, molding them together into a functioning unit with far greater resilience and adaptability than either organism has alone.

Other recent studies into planktonic life revealed plankton distribution patterns in time and space that had only been guessed at previously. Many phytoplankton are in fact drifters, incapable of swimming effectively; these float wherever the currents take them, collecting in large numbers wherever water masses of different temperatures and salinity meet. But other phytoplankton can swim actively, and many zooplankton cover large distances under their own power and aggregate of their own accord. Tightly packed swarms of surface-dwelling zooplankton can stretch for miles, while tiny, local clusters may pass completely unnoticed beneath the hulls of research vessels. Many open-ocean zooplankton make daily migrations from the surface of the sea to depths of nearly a mile. These journeys are thought to be an evolutionary effort to take advantage of the benefits of grazing on phytoplankton (which can live only in the photic zone), while reducing to some extent the threat of predation by plankton-eating fishes (which are also most abundant in upper water layers). These migratory animals are often so densely packed that they form what is called the deep scattering layer—a mixed group of zooplankton and fishes that give sonar systems a false image of a nearly solid surface hanging in midwater.

Zooplankton are also known to swarm actively but unaccountably at disparate times and places. Off Hawaii's Kona coast recently, millions of amphipod shrimp converged in several swarms in about 30 meters of water, clouding the water with their numbers, and swimming in all directions at random. Then, as if on some prearranged cue, thousands of the tiny shrimp gathered into amorphous aggregations that grew and lengthened into tornado-shaped funnels hovering in the water.

As the funnels grew still larger, small, ball-shaped clusters of animals appeared at the end of each funnel and began to drop successively toward the ocean floor like droplets of one liquid falling through another. At a depth of 3 to 5 meters, the falling balls disappeared as the amphipods dispersed into the water. Clusters continued dropping from the tornados until the animals became con-

Delicately constructed, single-celled foraminiferans actively prey on shrimplike copepods (Manfred Kage/© Peter Arnold Inc.).

centrated 5 to 7 meters below the surface. Then the animals aggregated again, rose to the surface in a bell-shaped swarm, and repeated the process.

Several hours later, all the amphipods were gone; the water was as crystal clear as it always had been. Local fishermen who ply those waters had never seen such a swarm before, and scientists in the area have never reported one since. If this occurrence, in some of the most heavily frequented waters in the tropics, had never even been seen before, imagine how many other aggregations of unknown size and purpose occur in the vast reaches of the open seas.

On a smaller scale, diving plankton biologists have found what they call micropatchiness throughout the photic zone as well. On bits of the amorphous organic matter called marine snow and on clusters of floating algal threads live aggregations of bacteria up to 10,000 times more concentrated than bacteria in the open water. Along with these bacteria, which include both primary producers and decomposers, drift hordes of their protozoan predators. Further study of these drifting particles has led to discoveries indicating that the microscopic communities they harbor may in fact be nearly complete ecosystems in miniature.

On each of these floating islands of life in a nutrient-poor sea, bacteria grow, respire, and are eaten, primary producers resorb any nutrients released in the process, and other grazers crop the primary producers. The whole system cycles constantly through intense microbial activity and with little input from the outside other than sunlight.

These floating microenvironments are seen by some biologists as highly evolved, stable associations of primary producers and consumers. By sticking close together, microscopic plants and animals can concentrate and store nutrients in their immediate vicinity in order to survive in the nutrient-poor macroenvironment that surrounds them. Chain-forming diatoms of several genera, for example, grow into clumps in which individual algal cells are separate but held together by a mucous sheath. That sheath houses a complete community of bacteria, diatoms, single-celled animals of several kinds, and even small, shrimplike zooplankton. The larger animals in these associations graze on bacteria, diatoms, and even on each other, forming a complete ecosystem bundled up in one tiny, floating package. The cycle of primary production, grazing, release of nutrients, and assimilation of those nutrients goes round and round, either on those tiny, floating particles or in small water patches too small to be sampled in any standard way.

Supporting the data on these microenvironments, other researchers have uncovered evidence that floating bacteria and other minute organisms are far more important as a group than anyone had previously imagined. These organisms are so small that they are called not simply nanoplankton, but ultrananoplankton. As tiny as 0.2 to 20 microns in size, bacteria and many single-celled protozoa such as flagellates and ciliates slip right through the mesh of standard plankton nets, leaving no way to estimate their numbers or ecological importance from plankton tows. Yet recent estimates show that microbial activity in the water column may in fact dominate the ocean's surface-layer food webs.

All these data, together with data on phytoplankton growth in laboratory cultures, lead a growing group of oceanographers to declare that the long-standing picture of the open ocean as a desert populated only by a slow-growing population of nutrient-starved plankton is in error. This controversial and fascinating work suggests that phytoplankton in the open sea may be growing just as quickly and be just as well-nourished as phytoplankton in more productive regions. What happens, say these oceanographers, is that nutrient scarcities limit the total amount of living tissue supported in the open sea without restricting the rate at which energy flows and nutrients cycle through the system. Instead of being represented as an ecosystem with a nearly static set of nutrient wheels, mired by lack of nitrogen and phosphorus, planktonic systems may be more accurately described as having wheels that are very small (carry few organisms at any time) but spin very rapidly.

In their self-contained microenvironments, individual organisms show few if any indications of being nutrient-limited. The difference between these systems and those of upwelling areas, then, is not in the organisms' absolute growth rates, but in the structure of the food chain and in the difference in total amount of living tissue supported as a result of input of new nutrients. The greater the nutrient input from outside the system, the more living tissue can be supported.

The highly evolved relationships among the members of these tiny floating worlds, like the examples of symbiosis between foraminiferans and dinoflagellates, provide miniature paradigms for interactions of larger organisms in larger systems. From planktonic microcosms to units as large as ecosystems—and, ultimately, to the biosphere in its entirety—assemblages of organisms take on a life of their own, above and beyond the lives of the organisms composing them. Growing and changing as its component creatures appear, reproduce, and die, each dynamic unit is a vital strand in the global web of life.

Even giant lobsters such as this one began their lives as tiny, free-swimming members of the plankton.

Capriccio on a Coral Reef

Corals

Arms waving in the current, a basket starfish captures plankton on the reef at night.

Half a world away from the frigid, productive Gulf of Maine and at the opposite end of the Pacific from the California coast's lush kelp forests, the Red Sea gleams like a jeweled finger pointing northward between Africa and Arabia. The desert air above the sea is crisp and dry; the naked mountains around it are pale orange and gray; the water filling it seems an impossibly vivid turquoise. The stark forms and pale colors of the terrestrial scene are unsettling to observers from more forested regions: no insects chirp, no songbirds sing, and no tree leaves rustle in the breeze. This is nature in one of her most desolate forms.

The ocean here contains little more in the way of nutrients than do the barren desert rocks surrounding it. There is no deep-water upwelling and no appreciable fresh-water runoff to provide fresh supplies of nitrogen. It is, in fact, the virtual absence of nutrients that gives the gulf its crystalline hue. No silt from fertile rivers clouds it, no blooms of phytoplankton stain it green.

Yet while the terrestrial desert stays barren, its aquatic counterpart teems with life. Beneath the water's surface, where bedrock plunges vertically into the depths, corals of all sizes, colors, textures, and shapes cover every centimeter of exposed rock. Hard corals, which build the reef by laying down skeletons of calcium carbonate, grow in the forms of massive boulders, tiny fingers, and huge branching elkhorns. Soft corals, which either have no rigid supports or depend on proteins rather than on calcium carbonate for their skeletons, form purple sea fans, brilliantly hued sea whips, and pastel-colored masses that undulate slowly in the gentle current.

Although their beauty might give the impression that they are merely decorative additions to the underwater landscape, these sedentary corals make possible the abundant life in these and other nutrient-poor tropical waters. The magic that supports this paradoxical fertility is a masterfully engineered symbiotic relationship between the soft-bodied animals called corals and single-celled zooxanthellae—members (in a special, vegetative stage) of a dinoflagellate species that swims actively in the waters surrounding the reef. Together they produce a network of living rocks upon which all the other animals depend for food, shelter, or both. Like mangrove roots and giant kelps, corals are both living members of their ecosystem and vital components of its physical structure. Their presence creates the complex, three-dimensional structures called coral reefs that are home to some of the sea's most bizarre, colorful, and fascinating creatures.

Coral organisms, *sensu strictu,* are animals, classified in the phylum *Cnidaria* along with jellyfishes and sea anemones. Living coral tissues

More than 50 meters beneath the surface, a diver examines a lone clump of gorgonian corals.

Following pages:
Sturdy yet flexible, the branches of a gorgonian colony space a myriad of eight-armed polyps to catch the prevailing current.

Pumped full of water and expanded to feed, these hard coral polyps will retract into their calcium carbonate skeletons at sunrise.

behave much as an inflatable water balloon filled with jelly would if stretched over deeply fissured rocks. By day, the balloon is deflated and the jelly dehydrated. The individual coral animals, or polyps, withdraw so deeply into the recesses of their protective calcareous skeletons that they virtually disappear. Each collection, or colony, of polyps looks to the untutored eye to be nothing more than an elaborate extension of the bedrock beneath it—some sort of multifingered stalagmite, perhaps.

It is only by night that most corals expose their animal natures. Polyps all over the colony pump themselves full of water and spread their tentacles in delicate coronas that, in many species, have the appearance of living snowflakes strewn in rows over the colony surface. Hidden in the fragile-looking arms of these snowflakes are stinging nematocysts that pierce, paralyze, and hold any small planktonic animals unfortunate enough to be swept against them.

That most corals use their tentacles to eat at least some plankton is unquestionable. But the total volume of plankton available to the reef is very small, and a number of calculations have shown that if corals were simply animal planktivores they could not survive on the plankton stocks available in these nutrient-poor waters. Nevertheless, a dense aggregation of animals stubbornly insists on building a food web on a seemingly nonexistent base—a puzzle to generations of biologists from Darwin's time onward.

While many questions remain, recent research has come close to explaining how the living reef obtains its necessary nutrients. The key to the reef's productivity is the closely knit symbiotic relationship between coral animals and the thousands upon thousands of zooxanthellae that thrive inside the body walls of hard coral polyps. Zooxanthellae and coral animals together form a short-circuit ecosystem, much like those found in planktonic foraminiferans, in which primary producer and animal are so intimately connected that they behave as a single organism.

As is the case with foraminiferans, coral animals generate carbon dioxide and ammonia, wastes for the coral animals but perfect fertilizers for zooxanthellae, which absorb the compounds directly from the animal tissue around them with minimal effort. So efficient is this system that several coral species release virtually no nitrogen to the surrounding water at any time, indicating that their algal symbionts absorb ammonia twenty-four hours a day.

Coral animals make their single-celled guests as comfortable as possible. In addition to providing them with nutrients, coral polyps position their algae in the proper locations to receive maximum doses of valuable sunlight. During sunlit hours, the polyps' feeding tentacles are withdrawn and their tissue layer containing zooxanthellae is spread out like a carpet to catch as much sunlight as possible. In the dark, when photosynthesis is no longer possible, the algae are withdrawn and the tentacles expanded. Corals that grow in shallow water under intense sunlight pile their algae up in numerous layers and grow in branching colonies to pack in as many algal cells per unit volume as possible. In deeper water, where less light is available, the coral colonies flatten out horizontally, forming tables over which each animal spreads its photosynthetic carpet as carefully as any farmer might lay out a field of corn.

Zooxanthellae donate a large proportion of their photosynthetic output to their hosts. Like all plants, these algae generate oxygen as they photosynthesize; and given near-optimum supplies of light and nutrients for a good part of the day, zooxanthellae steadily photosynthesize at high rates. Most of the oxygen they produce is directly absorbed by the animal tissues, a sort of quid pro quo for the carbon dioxide they absorb from their hosts.

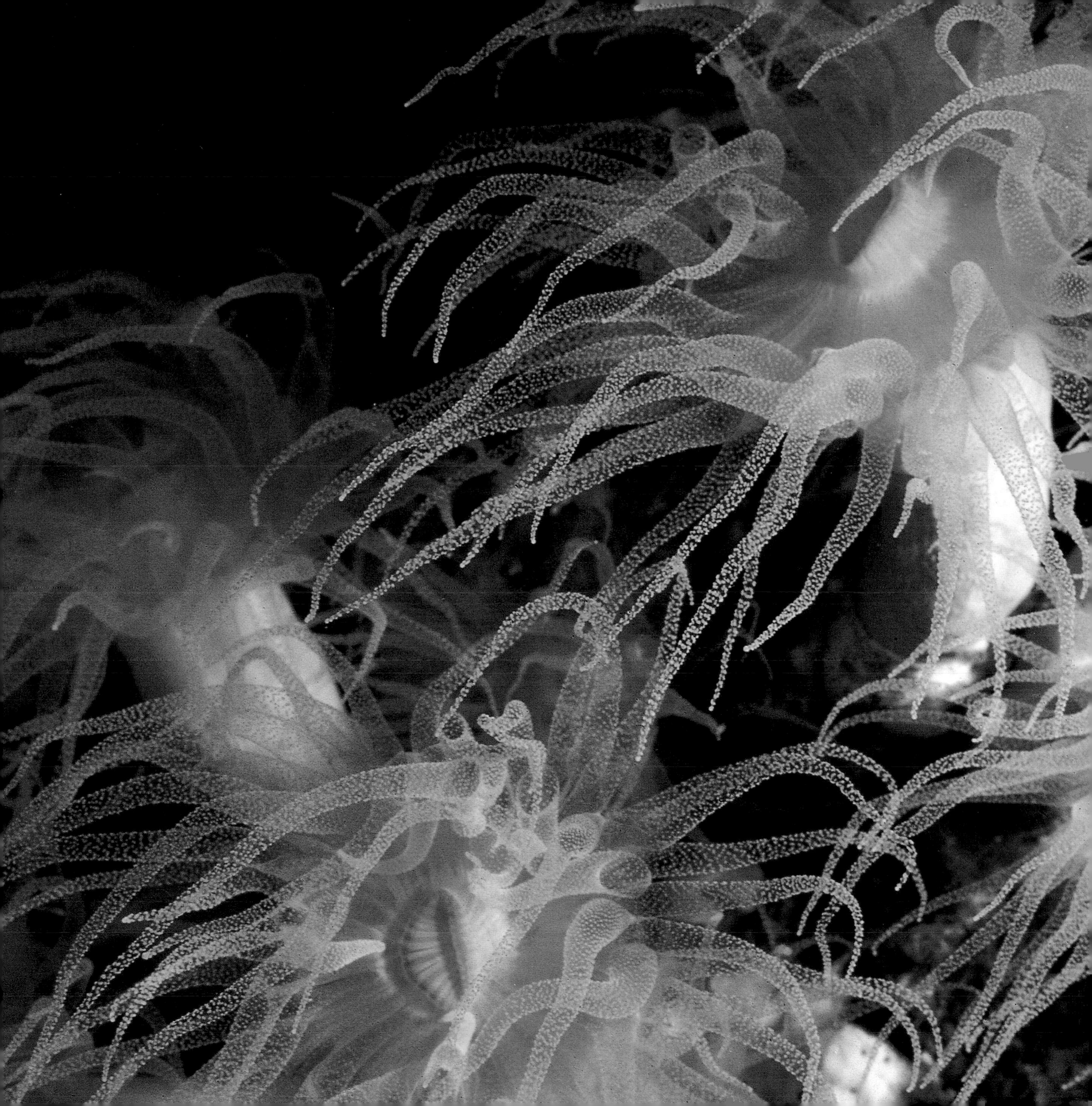

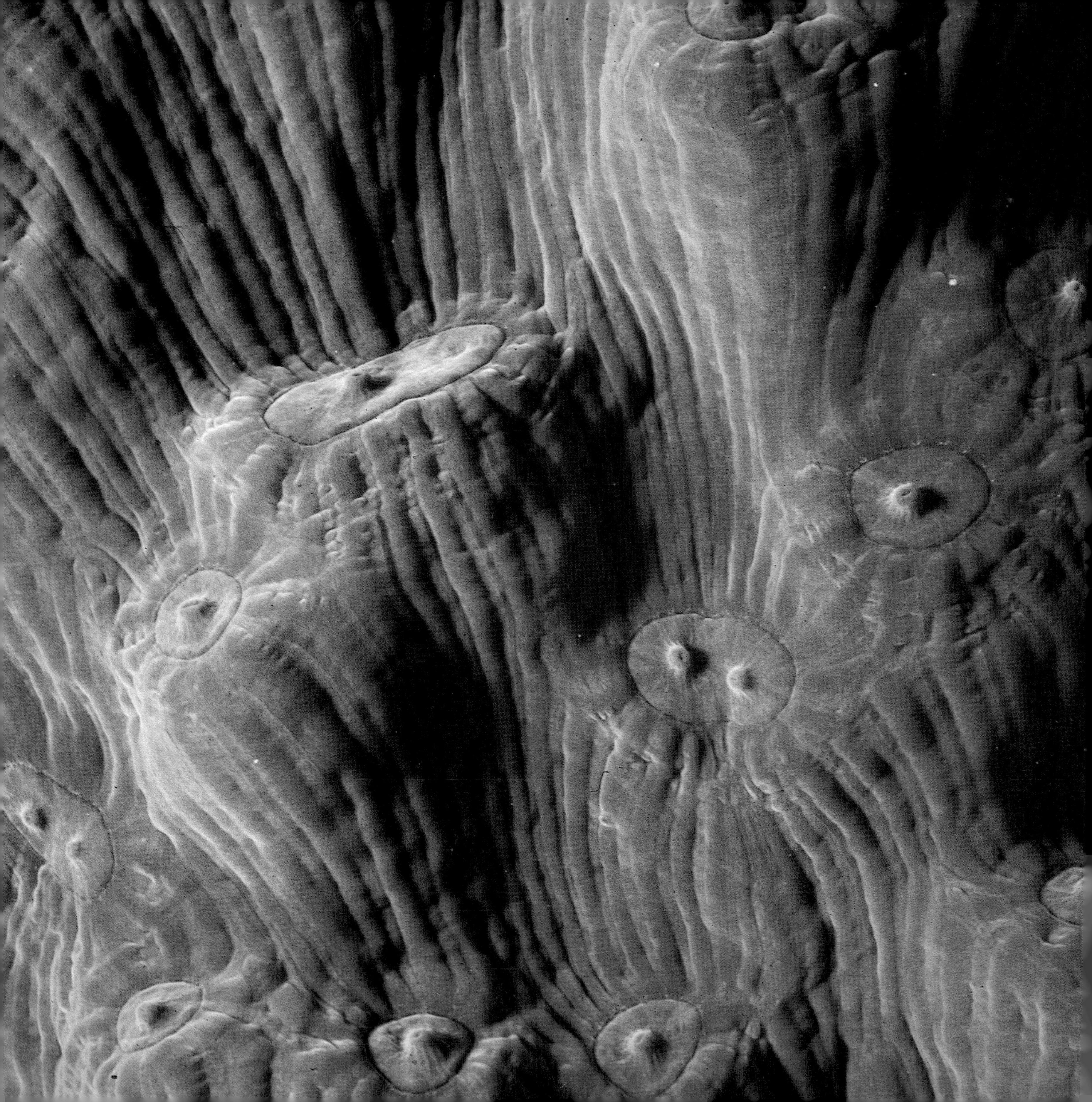

But the symbiosis does not stop here. Zooxanthellae also manufacture a suite of sugars, amino acids, and other organic compounds that are absorbed directly by the coral tissues. Some compounds may even be passed back and forth between animal and plant two or three times, being altered by or combining with other molecules each time. Though no one yet understands how this relationship is regulated, the algae clearly provide their hosts with compounds that could just as easily serve algal growth. The release of the compounds is stimulated by the coral tissue, but the nature of the stimulus is unclear.

Another mystery is the manner in which zooxanthellae help hard corals manufacture and deposit their calcareous skeletons. Whether deposition of calcium carbonate is directly stimulated by organic molecules that the algae give off or whether deposition is inhibited by waste products the algae remove from coral tissues is still not known. In any case, careful experiments have shown unequivocally that, without the constant assistance of their zooxanthellae, reef-building corals simply cannot grow.

Therefore, biologists view hard coral colonies as a remarkable combination of plant and animal that in the course of twenty-four hours acts as primary producer (by photosynthesizing), as herbivore (by consuming products of photosynthesis), and as carnivore (by capturing plankton). When properly operational, the coral-algal symbiosis circumvents parts of the great energy problem intrinsic in ordinary food webs. Normally, the accumulation of nutrients by one animal from another inevitably involves a loss of energy; here, newly ingested nutrients can be combined with freshly captured solar energy in a highly efficient fashion. This extraordinary partnership does, however, require the active participation of both plant and animal functions to keep the colony operational. Starve well-lighted coral colonies, and they will lose weight. Feed coral colonies in darkness, and—even though they devour the zooplankton offered to them—they will fail to grow.

The importance of zooxanthellae to the growth of reef-building corals can be seen clearly. At increasing depths, many shallow-water corals disappear, and others change shape dramatically. At depths where light begins to dim, zooxanthellae in finely branched, fingerlike colonies cannot capture enough energy to photosynthesize, and corals with simpler, rounded forms increase in frequency. Still deeper down, even the kinds of adaptations to lower light intensities used by deep-water algae cannot maintain the high rates of photosynthesis necessary to maintain vigorous colony growth. Here, some species that form boulderlike colonies in brighter locations grow into flattened table shapes that require less skeleton to support the same amount of living tissue. Simultaneously, these flattened colonies stretch out horizontally from the reef wall to expose their algal symbionts to as much light as possible.

Invisible adjustments to changes in light intensity accompany these obvious changes in colony shape. Zooxanthellae in corals in dimly lit locations carry up to four times as much chlorophyll per cell as those living in brighter light. Coral animals in deep water must capture the same quantity of zooplankton as shallow-water animals of the same species, but they must get by with substantially less help from their zooxanthellae. And among the many genera and hundreds of species of corals, each has evolved a different way of balancing feeding, photosynthesis, and growth in different locations.

Even with these changes in physiology and colony structure, however, reef-building coral colonies reach their limits at depths of about 60 meters. At depths below this, light is simply too dim for zooxanthellae to grow well. Larvae of reef-building corals may well settle in deeper water,

By day, the polyps of reef-building corals withdraw tightly into the recesses of their colony's skeleton, while zooxanthellae photosynthesize rapidly in the tropical sun.

and they may even be able to survive for a time by plankton feeding, but they ultimately find it difficult if not impossible to build their calcareous skeletons.

As if functioning simultaneously on three trophic levels were not enough, many coral species have evolved still other nutrition-gleaning techniques. Some ingest substantial quantities of detritus and bacteria by casting out and retrieving sticky strands and nets of mucus. Several species are covered with countless tiny, fingerlike projections called microvilli, similar to the larger projections found inside human intestines. Just as intestinal linings absorb food from the human gut, corals' microvilli probably help them to take up scarce dissolved organic nutrients directly from sea water.

Soft corals that never make calcareous skeletons often lack zooxanthellae and probably obtain many of the nutrients they need in these alternate ways. One subclass of soft corals, the alcyonarians, includes a number of species that form delicately branched colonies having the consistency of day-old Jello, in shades of green, white, yellow, orange, red, and scarlet. Horny corals, or Gorgonians, build skeletons composed of proteins similar to those in human hair and fingernails.

Because they do not depend on algal symbionts, many soft corals, along with cave- and deep-dwelling Gorgonian species, are not restricted to sunlit locations. Although they are often outcompeted by the reef-building corals in open, shallow water, soft corals manage better in dimly lit locations such as caves, overhanging ledges, and the deeper sea bottom.

Where corals grow, hundreds of other organisms congregate. Their combined nature as living tissue and solid structure enables corals to contribute to their environments in many ways. Directly and indirectly, these multitrophic-level creatures provide food for a host of organisms. Some coral feeders, including planktonic bacteria, crabs, shrimps, mussels, fishes, and even other corals, depend on abundant coral mucus for such energy-rich compounds as triglycerides and fatty acids. Some planktonic shrimps can live almost entirely on a diet of coral mucus. Other animals, among them certain starfishes, urchins, fishes, and crabs, eat living coral tissue itself.

Equally important as corals' position in food webs, however, is the manner in which their colonies provide shelter for resident fish and invertebrate species. By building the undersea metropolis that is a coral reef, corals set the stage for an evolutionary drama in which they serve both as the stage itself and as active players.

At depths of over 75 meters, hard corals virtually disappear, to be replaced by deep-water Gorgonians that lack zooxanthellae.

The Splendid Kingdom

Following pages:
Tunnels honeycomb the reef, providing shelter for scores of fishes and invertebrates.

Viewed from a distance, coral reefs are among the most awe-inspiring creations of life on earth. The Great Barrier Reef, to choose one example, stretches along Australia's coast for almost 2,000 kilometers and contains in the vicinity of 21,000 cubic kilometers of limestone rock, all assembled by countless tiny polyps over many centuries. The magnitude of this structure is almost inconceivable; the bulk of even the greatest megalopolis constructed by human beings is dwarfed by comparison.

Close up, reefs are equally astonishing. Their complex, brightly colored colonies of hard and soft corals support and form the backdrop for a collection of beautiful and bizarre creatures whose eclectic assemblage of colors, shapes, and behaviors crowds the outer limits of imagination. Branches, boulders, fingers, antlers, fans, whips, and tables of coral reach out and across each other like the turrets and domes of the pastel-colored skyscrapers that grace the covers of science-fiction novels. Labyrinthine tunnels penetrate far beneath the surface, and everywhere fishes and invertebrates swarm.

Living systems as grand and complex as this exceed our present ability to understand in their entirety. The truths about the macrocosm of the coral reef—like the truths about global ecology—exist in some multidimensional space beyond our ability to conceptualize. Studying such a vast collection of phenomena requires researchers to limit themselves to one or two questions at a time and to rely on one or two specific modes of investigation. Slicing into multidimensional reality in this fashion bares for inspection a two- or possibly three-dimensional view of that reality, but it never reveals the entire story. Only by carefully selecting the mental cutlery, judiciously choosing the angle of each slice, and then painstakingly reconstructing the whole by comparing the views afforded by each slice can scientists ultimately assemble a working approximation of the entire system.

Perhaps the best example of science's ability to describe reef life without being able to explain it can be seen in the effort to address the staggering diversity of animals and plants packed into this ecosystem. Coral reefs in the western Pacific shelter more species than almost any other habitat on earth. This warm-climate phenomenon is well known and has terrestrial parallels; just as tropical reefs have more species than salt marshes, tropical rain forests make even the lushest temperate-zone woods seem species-poor by comparison.

Off-the-cuff justifications for this diversity are often substituted for real explanations of how it evolved. Some writers note that reefs occur only in the tropics and that diversity generally increases in habitats from the poles to the equator. Others point out that intense sunlight is available year-round in the tropics and provides a constant source of energy to fuel high primary productivity. Both these facts are indisputable observations of how reef habitats might allow for high diversity, but neither really addresses the issue of how that diversity arose or how the reef system maintains that diversity today.

In fact, no single explanation yet proposed seems able to explain reef diversity in its entirety. There are, however, several intriguing facts and theoretical perspectives regarding reef animals—slices of that frustratingly complex reality—that have been revealed by different approaches to the riddles of reef life.

Among these are observations that the coral reef is one of the oldest ecological communities preserved in the fossil record. The first reeflike assemblages of animals and plants appeared more than 600 million years ago; relatives of modern reef-building corals date as far back as 200 million years. Reefs have been evolving—with several ma-

jor interruptions—ever since then, as continents jockeyed into their present positions and glaciers came and went. Even after major periods of extinction in the sea that wiped out thousands of species, reeflike systems have repeatedly reassembled from surviving groups of earlier reef dwellers. Reef organisms, in contrast to temperate estuarine species, have thus had plenty of time to evolve complex interrelationships, both with their physical environment and with their neighbors.

Then, too, life in many tropical seas not only receives the constant high level of solar energy noted above, but also benefits from freedom from freezing temperatures, a relatively constant level of salinity, and stability in the nutrient supply. Instead of evolving structural and physiological adaptations to the environmental extremes encountered in temperate and polar zones (theorists such as Darwin's contemporary A. R. Wallace have pointed out), animals in stable, tropical environments can use their genetic variability to evolve specialized responses to both biological and physical features of their surroundings.

In fact, reigning ecological and evolutionary theory predicts that, under stable, predictable environmental conditions, high-diversity communities not only can evolve but should evolve. This prediction arises from an ecological theory of competition that has its origin in the writings of Malthus and Darwin. Given the struggle for existence and the inevitable shortages of limiting resources, competition over resources is unavoidable. Such competition may be between similar species or between populations of a single species, and it allows no two groups of organisms to exploit in a stable fashion exactly the same resources in exactly the same place at exactly the same time. Chances are that one group will be more efficient at what it does than the other and will have better success at survival and reproduction.

One possible result of this imbalance is the extinction of the inferior competitor, a process ecologists call competitive exclusion. Under predictable environmental conditions, however, there is an alternative: two groups of organisms competing for the same resource could—through the accumulation of random genetic variation that underlies all evolutionary change—evolve behavioral and morphological specializations that permit more efficient use of a particular subset of resources.

Those individuals that became more efficient at using resources not in demand by other individuals would experience less competition and would be rewarded with higher reproductive success. As long as a stable environment ensures availability of a variety of similar yet distinct resources, natural selection will favor this process of subdividing. What was once a single niche will become two or more smaller niches that overlap only minimally—a process called resource partitioning.

Under the right circumstances, competitively based evolution of this sort can cause two isolated populations of a single species to split into two distinct species. Under other circumstances, it could cause two ecologically similar species to diverge in their ecological characteristics. An entire generation of ecologists has come up with evidence that evolutionary pressure of this sort is the guiding force behind the evolution of the varied strategies organisms employ to divide up food and space on the reef.

Plankton feeders, supported by a dependable food supply, divide up their potential food species in nearly every conceivable fashion. Planktivorous damselfishes stay close to the reef. *Anthias* swarm after plankton in the water above the reef. Squirrelfish hunt for plankton only at night. Still other species leave the reef altogether to feed on open-water plankton—some during the daytime, others only after dark.

Triggerfish, strong swimmers with powerful jaws, feed on small crustaceans and echinoderms both on and around the reef.

The morphology, feeding behavior, and preferred territory of each species are just a little different from another's. Certain butterflyfish eat nothing but the living polyps of one particular coral species; others dine on nothing but certain sponges. Still others, sporting jaws that look like needle-nosed pliers, reach deep into cracks and crevices for hidden invertebrates. Even herbivores, such as many tang species, each graze only a few select species of algae. Triggerfish use powerful beaks and adaptable predatory behavior to make short work of even heavily armored sea urchins: carefully lifting an urchin by a few of its spines and dashing it against rocks, a clever triggerfish can break the remaining spines until the urchin's tasty soft tissues are unprotected. Alternatively, a triggerfish may use its mouth to direct powerful water jets at an urchin, blowing it over and exposing its unprotected underside.

According to the stability-and-competition hypothesis, predictability is essential, not only in creating this ecological diversity, but in maintaining it. Species that are ecological generalists can often withstand environmental changes that cause shifts in their food supply; they simply change their feeding behavior to exploit a different part of their broad niche. Specialized species often lack that adaptability; if the food on which they have come to depend vanishes—even temporarily—the specialization that proved so valuable can turn into their evolutionary Waterloo.

Environmental stability clearly does play a role in the life of modern reef communities. Reefs off the southern coast of Florida, where water temperatures periodically dip too low for most corals' comfort, are less diverse than those in perpetually warm climates. Off the Galápagos Islands, periodic changes in winds and currents can cause temporary upwellings of cool, nutrient-rich bottom water in certain locations. Although these unstable, unpredictable conditions prove a boon to attached algae and phytoplankton, they severely restrict and in some cases eliminate reef growth. More stable areas off nearby islands have more diverse and well-established reef communities. Similarly, other work has shown that Red Sea coral communities are less diverse in shallow areas—areas more exposed to the disturbing effects of wave action, tides, and changes in air temperature—than in slightly deeper water.

The stability-and-competition hypothesis was accepted nearly universally for years as the basis for extensive species differentiation and coexistence, and studies still appear monthly in scientific journals showing how some group of reef animals divides up one or another available resource. And yet evidence is accumulating that stubbornly refuses to be explained by this long-approved argument.

Some recent studies show that areas with seasonal upwellings and temperature variations support more species than nearby, more constant environments. At the same time, ichthyologists point out that, although different reef fishes do exploit different resources, groups of species (called guilds) exist whose members are not ecologically very different from one another.

Each guild contains species that are fairly specialized as fishes go; reef communities support guilds of plankton feeders, guilds of herbivores, and guilds of coral eaters, among others. Yet within each guild, individual species are often fairly similar. Several damselfish members of the plankton-feeding guild, for example, seem to differ very little in their ecological requirements. Experiments show that determination of which particular member species within the guild comes to live in a particular spot depends more on chance than on any predictable ability to compete for food or space. Thus, theories of reef ecology must broaden to allow unpredictability and the luck of the draw to usurp a bit of power from competition.

Long-nosed hawkfish, usually found among deep-water Gorgonians, are rarely sighted at depths shallower than 25 meters.

Many scientists have begun to suspect that the reigning doctrine of stability, competition, and resource partitioning has been applied both too broadly and too uncritically. It is an axiom of scientific investigations in complex systems that if researchers enter their studies with particular beliefs or expectations, they will (consciously or unconsciously) search most actively for data that confirm those beliefs or expectations. This is not to say that honest researchers consciously falsify results; it merely points out that many do not seek hard and long for data that foster alternative hypotheses. Once competition was crowned king of the theories explaining community structure, these modern critics claim, true objectivity was lost. Students whose work turned up no evidence of competition were told to go back and look again until they found it. Not surprisingly, most did.

Although many studies of reef habitats do show differences in jaw size or in feeding habits among similar species, no one has satisfactorily shown how different two species must be in order to avoid competing with each other. Nor have researchers produced an experimentally verifiable rule describing how much ecological overlap is permissible before interspecific competition becomes severe enough either to drive one of the competitors into extinction or to force further evolutionary changes between competitors. Without a specific prediction that can be experimentally proven or falsified, two researchers with differing views on the importance of competition can look at the same data and come up with opposite conclusions on whether or not the species in question are competing.

Another problem with the stability-and-competition hypothesis is that there is no hard and fast definition of how "stable, predictable" environments differ from "unstable, unpredictable" ones. In fact, while advocates of competition have been touting the reef as stable and predictable, an entirely different crop of ecologists have proposed new hypotheses that emphasize just how unpredictable and unstable the reef environment really is for its inhabitants!

These biologists do not dispute the fact that the reef is blessed with long-term, large-scale stability. They point out, however, that the stability of the reef as a whole means little to individual reef dwellers that interact with their environment on a much smaller time scale and in relatively restricted areas. Many reef organisms, ranging from algae through corals to worms and even fishes, either spend their entire adult lives physically attached to a single spot or behaviorally tethered to a limited territory. For this reason, short-term, local changes in reef structure and biology can mean a great deal to the survival of these organisms.

Small-scale changes are indeed plentiful on the reef. Storms can wipe out whole stretches of living coral, either by wave action or by freshwater runoff. Worms, sponges, and other organisms constantly bore, chip, and eat away at the bases of hard coral structures, periodically causing them to collapse and tumble down the reef. Coral colonies themselves contribute to the constantly changing face of the reef, both by building new colonies and by competing with each other for what is apparently the major limiting factor for hard corals in many locations—space on hard substrates.

In battles whose intensity surprises those who think of corals as inoffensive, sedentary creatures, coral polyps use their nematocyst-loaded tentacles to fend off sponges, worms, the larvae of other corals, and algae—preventing the newcomers from settling on top of them. Many corals, especially soft corals, have developed their own chemical bactericides, algicides, and fungicides to help keep them free of the growths that quickly cover any unprotected hard surface on the reef. Where one

Following pages:
The complex, three-dimensional structure of the reef forms a continually changing patchwork background for reef dwellers.

living coral colony grows, therefore, other sedentary organisms usually cannot.

Corals also stage active battles with their neighbors over growing space. Some wage physical warfare on any other coral colony that grows too close by extruding long filaments that kill and digest the interloper's thin skin of living tissue. When two species battle each other in this way, the winner can usually be predicted, as a definite digestive hierarchy exists among reef-building corals; the victor in such confrontations then proceeds to overgrow the vanquished. With dozens of coral species competing on a reef, however, the long-term victors may be impossible to predict. A species that might be destroyed in direct digestive warfare may be fast-growing enough to spread up and over its neighbor, shade its zooxanthellae, deflect food-bearing currents, and ultimately kill it without physical contact. Some species exude chemicals toxic to other coral species, inhibiting their growth before the two colonies get close enough to engage in digestive warfare. Some soft coral colonies use their fleshy colony bases to "walk" across hard coral colonies, smothering them and leaving a dead, open, coral rock trail in their wake.

All these interactions occur more or less unpredictably, creating random changes and openings on the reef surface. In the same spot where one species of coral flourished two years ago and another species grew last summer, a patch of green or coralline algae might be thriving now. Thus, although the reef as a whole may experience stable physical conditions, its surface—far from affording its inhabitants a stable, living framework—much more nearly resembles a continually changing mosaic.

To organisms that prefer to live on particular sorts of patches on the reef, this continual shifting has two consequences. On the one hand, it decreases the probability that any single species will find conditions favorable enough for long enough that it can take over the whole area and exclude its competitors. On the other hand, it constantly opens up new areas for colonization, and makes each of these randomly opened areas available for settlement by whichever suitable species happens to get there first. Because most reef organisms—from algae to fishes—produce large numbers of free-swimming larvae, just which species of coral or damselfish gets to an open area first may depend more on the luck of the draw than anything else. Both of these phenomena can prevent competition from excluding species that might prove to be competitively inferior under more constant conditions.

Thus, the very data traditionally used to assert that reef diversity is due to stability, predictability, and competition can also sustain a hypothesis asserting the importance of unpredictable local changes and of chance in allowing the continued survival of so many species. This view has been dubbed the "chaos hypothesis."

Still another hypothesis attempts to combine both order and chaos by proposing that the guild—not the species—is the ecologically significant unit on the reef. According to this hypothesis, the processes outlined under the stability-and-competition hypothesis have led to the evolution of the ecologically diverse guilds present on the reef. The luck-of-the-draw mechanisms proposed by the chaos hypothesis, on the other hand, explain how several ecologically similar species within each guild can coexist without driving the others into extinction. (A logical extension of the chaos hypothesis might propose that species within a guild evolved their limited differences more or less randomly, rather than in response to competition.)

Operating within and around these hypotheses are observations that emphasize interactions other than competition between species. Just as the nor-

Seen here in a Hawaiian lava cave, trumpetfish are twilight-active hunters.

mal feeding habits of sea otters can maintain the integrity of a kelp forest, the daily activities of predators, parasites, and symbionts can maintain diversity on the reef. Many herbivorous species like tangs and surgeonfish, for example, form huge schools that browse across the reef from sunup to sundown. Lacking some of the digestive enzymes needed to digest their algal food efficiently, they must eat continually to fulfill their nutritional and energy requirements. As a result, these species devour a tremendous quantity of quick-growing algae that, if left unchecked, would take over many reef areas normally occupied by corals, sponges, and other organisms with which the algae compete for space.

Other species, including several very common damselfishes, affect the reef face by establishing territories within which they kill living coral tissue and from which they chase other plant eaters. The result, in the middle of a live coral patch, is a lush algal lawn. Within the protected confines of this jealously guarded lawn flourish several small invertebrates that are rare elsewhere, thus increasing the local diversity of these species.

But the importance of this interaction goes much further. As damselfish kill living coral tissue, they expose portions of the coralline skeleton beneath. This opens a path for coral-boring organisms, including worms and sponges, that ordinarily cannot pass the defenses of living coral polyps. Undermined in this manner, coral colonies inside damselfish territories are more likely than others to be damaged by large fish or storm waves, resulting in further disturbances on the reef. Then, too, because damselfish prefer some coral species to others, their presence in an area can directly affect which coral species is dominant there.

Then, too, the activities of all coral predators contribute to shaping the living reef. The crown of thorns, a starfish, can devour entire coral colonies; it also seems to prefer some species over others. When these starfish increase in numbers explosively, as they did in the 1960s and again in the early 1980s, they can radically change the face of the reef. At a less cataclysmic pace, coral-eating butterflyfish nibbling on individual polyps and predatory sea urchins grazing either on attached algae or on the exposed tips of growing corals may also tip the precarious balance among competing coral populations in one direction or another.

As much as coral-reef animals affect coral growth and diversity, the complexity of the interlocking reef structure built by the corals is itself a major contributor to the diversity of coral-related animals. A vast network of tunnels and caves honeycombs the reef structure. Some caves are little more than pockets in the living coral wall; others wind for many meters through the bowels of the reef. Some are shallow and sunlit; others, deep within the reef, remain perpetually dark. Caves can be as barren as terrestrial deserts, or they may teem with rare and unusual organisms.

The deepest, darkest, and most barren caves trace their origin back to the ice age, when glaciers tied up much of the world's water supply. Globally, sea level dropped many meters, stranding ancient coral formations well above the old water line. Over the centuries, slightly acid rain water percolated through these porous calcareous reefs, dissolving globe-shaped cavities in the rock and furnishing the chambers with delicate stalactites and stalagmites. When nature's caprice melted the glaciers and filled the seas once more, the rising waters flooded the caves and submerged them far beneath the surface. Deepwater corals and sponges grow outside the entrances of these dead-end grottoes today, but their interiors, cut off from life-giving light and food-bearing currents, remain as lifeless as the pharaohs' tombs.

Quite a different situation exists in shallow water. Here, stony corals build free-form colonies that grow over, onto, and around each other at random. As individual colonies die, skeletons left behind are quickly covered by coralline algae and cemented into a rigid matrix. Simultaneously, sponges and a variety of worms, snails, and bivalves attack any exposed rock, dissolving parts of the surface and riddling the interior with bore holes. Whenever these fragile, haphazardly assembled structures collapse during storms, small caves are formed. A series of local collapses often leaves large hollows connected by a maze of small tunnels that meander through the reef.

Still another type of cave appears suddenly when earthquakes open fissures in the bedrock underlying the living reef. Industrious coral colonies quickly roof the newly created chasm, producing tunnels that penetrate deep into the reef and run for many meters.

Coral caves are difficult growth habitats for many sedentary reef dwellers. Some organisms, such as algae and hard corals, need sunlight in order to survive and are excluded from all but the outermost meter or so near the cave's entrance—except for the rare locations into which shafts of light from the midday sun penetrate. Other organisms, which may be able to survive in the dark, either cannot tolerate the rain of silt that falls on cave surfaces or require steady currents to deliver plankton. The stillness of dead-end caves, combined with their total darkness, makes life there nearly impossible for most organisms.

The exclusion of many plants and animals does, however, open opportunities to certain odd beasts that can tolerate cave conditions. Space on hard surfaces is always at a premium on the reef; the competition for a firm place in the sun is often fiercer than competition for food. Because hard corals and most algae—which monopolize much of the space outside—cannot survive in cave conditions, sea fans, sea whips, and soft corals, found elsewhere only in the darkness of the reef's lower limits, can successfully grab holds near cave entrances where some current flows. Deeper within, cave walls provide a haven for wire corals, delicate hydroids, and sponges that also cannot compete for space outside. The seemingly barren, silt-covered floors shelter worms and molluscs that would be torn apart in minutes by fishes on the open reef.

But the most interesting function of coral caves that remain open to some light and circulation is the essential refuge they provide to fishes and mobile invertebrates during periods of required rest or inactivity. Many reef animals are active only by day, and must seek shelter by night; others emerge only after dark and hide from sunrise to sunset. So regularly do these animals alternate occupancy in coral caves that the reef bears comparison to a collection of boardinghouses whose rooms are rented to two sets of occupants, with the night guests and the day sleepers passing each other in the halls at dawn and dusk.

The Reef by Night

Where strong water currents flow through reef tunnels, white-tipped reef sharks are often found "sleeping."

Motionless in the evening gloom, a parrotfish stares out of its lidless eyes, enveloped by a transparent mucous cocoon it has spun under the protection of a coral ledge. Normally too nervous to allow divers within a meter and a half of it, this one seems oblivious to any disturbance. Contrary to appearances, the fish is neither ill nor injured. Several hours after sunset, it—like all the other parrotfish on the reef—is simply asleep.

Elsewhere, in innumerable cracks, crevices, and caves, thousands of brightly colored fishes that swarm around the reef by day have secreted themselves against the uncertainties of darkness. Gone are the clouds of *Anthias* that school in shallow water from sunup to sundown. Gone, too, are the angelfishes, butterflyfishes, and damselfishes that patrol their territories throughout the daylight hours.

But the nighttime reef is far from inactive. Invertebrates whose family ancestry dates back to the dawn of multicellular life in the sea crawl among the coral boulders or perch in feeding positions atop coral ledges. Fishes probe the darkness with living flashlights located conveniently beneath their eyes. Luminescent plankton sparkle with pale, blue-green light. These phantoms of the night are never seen by daytime divers because by day they make themselves as scarce as diurnal species do after dark.

The changeover from daytime to nighttime activity on the reef begins in the late afternoon. As the setting sun casts lengthening shadows onto the sea, daytime fishes alter their behavior in subtle ways. Black and white damselfish and silver-blue *Chromis,* which spend the sunlit hours high in the water column searching for plankton, gradually sink closer and closer to the reef. The complex, daytime body markings on butterflyfishes begin changing to darker, simpler patterns; and the animals themselves become increasingly skittish and draw ever closer to the familiar cracks and small caves of the reef. As twilight approaches, species after species—even normally aggressive ones—grow shy and nervous enough to flee upon the slightest provocation, like children frightened of things that go bump in the night.

The twilight discomfiture of diurnal fishes is well-founded. Many are relatively recent additions to the reef fauna and owe their places in it to the sharp eyes, good color vision, and novel hunting techniques they employed in carving out their niches. By day, their eyes function extraordinarily well, but at dusk they have become unreliable and by night utterly useless.

For prospective prey, twilight is a time of extreme danger. Just as human eyes begin to strain in the gathering gloom, snappers, jacks, lionfish, and sea basses swim boldly into view in search of prey. Moray eels, whose beady eyes and fanglike teeth rival those of any imaginary sea monster, slither from their tunnels beneath the reef surface. Sharks, relying as much on their phenomenal chemical and electrical senses as on their eyes, patrol the reefs with almost mechanical efficiency. Deceptively fragile-looking lionfish—which spend the better part of the day resting quietly among convenient shadows—glide gracefully around the coral ledges, searching for unwary, bite-sized prey. If it finds any small fish sufficiently nightblind or disoriented, the lionfish slowly but surely herds it into a dead-end alley with its fanlike pectoral fins and swallows it in a lightning-fast gulp.

These dusk- and dawn-active, or crepuscular, predators have evolved feeding patterns that take maximum advantage of the severe visual disadvantage their diurnal prey suffer in this twilight world of dim shapes and confusing silhouettes. The predators' own eyes have evolved specifically to function in dim light, enabling them to navigate and stalk their prey with ease. Although many

After a long day foraging for algae on the reef and in seagrass beds, a parrotfish shelters for the night on a ledge at the entrance to a coral cave.

Preceding pages:
At twilight, schools of *Anthias* descend en masse onto a large coral head where each individual wedges itself into a crevice for the night.

To mask its scent from nocturnal hunters, this parrotfish surrounds itself with a mucus cocoon.

of these predators will try to gulp down a quick snack almost any time the opportunity presents itself, their success is limited except in semidarkness, for by day their sharp-eyed prey easily outmaneuver them.

Threatened by these twilight hunters, and possessing neither sufficient means of defense nor senses capable of detecting predators in time, most daytime fishes have taken the evolutionary recourse of protective retirement. Members of each species, according to size and shape, squeeze into sheltering nooks and crannies among the coral formations—and sleep. So well do they sequester themselves that a novice diver could easily swim along the entire reef and imagine that all the diurnal fishes had simply vanished. By swimming slowly and peering carefully between coral branches, though, a careful observer can see fish after fish holed up for the night.

Anthias pack every space between the branches of finger coral colonies. From larger crevices protrude the tails of triggerfish that have locked themselves tightly into the reef by means of rigid spines that make it impossible to pull them from their hiding places. Only the fish themselves can release those spines, by depressing another spine—the trigger from which they take their name. Many larger coral ledges are shared by full-grown parrotfishes, whose mucous cocoons seem to protect them from predation, and goatfish, whose long, tastebud-laden barbels are now tucked neatly under their chins. Small wrasses take the protective step of burying themselves out of sight in sandy areas before sleeping.

As the twilight deepens further, an eerie calm settles over the reef, as even the crepuscular predators find it difficult to see. For about twenty minutes, an interval known as the "quiet period," there is hardly an animal to be seen. The daytime fishes are all tucked away, and the nighttime animals have yet to appear.

Then, slowly, the real creatures of the night leave their daytime hiding places. Schools of silvery fish called *Pempheris,* which spend the daylight hours in shadowy coral caves, emerge a few at a time during twilight to swarm along the reef wall. As darkness falls, swarms as large as several thousands mill around together before dividing into groups of one or two hundred. The groups then disperse in all directions into the open water, where they feed throughout the night. Predatory cone shells emerge to glide over the reef surface, searching for crustaceans and small, sleeping fish to stun with their poison-coated harpoons. Hermit crabs as large as a man's fist, their recycled gastropod-shell homes covered with hitch-hiking sea anemones, scuttle about with seemingly limitless energy.

The fall of darkness is like a journey back in time, for many specifically nocturnal reef animals are far older than dinosaurs. Among the most ancient are several members of the phylum *Echinodermata,* including the starfishes and the feathery sea lilies, or crinoids, whose fossilized relatives first appeared over 250 million years ago. Modern crinoids spend the daylight hours hiding in the reef, curled tightly into balls about the size of a small child's fist. As darkness falls, they uncurl their colorful, featherlike arms (a single animal can have as many as fifty-five of these) and crawl across the reef with a series of movements that somehow manage to appear simultaneously ponderous and graceful. Then, perching on the edge of the reef at a spot where the current meets their needs, crinoids gather in groups and spread their arms into semicircular fans set at right angles to the current. Unless disturbed by a diver's light, they remain in these positions until dawn, filtering plankton from the water as members of their class have done for millennia.

Joining crinoids in the quest for plankton are the writhing masses of tentacles belonging to basket stars. The Latin name *Gorgonocephalus* ("gorgon head") suits these nocturnal feeders well, for their constantly twisting tentacles look like nothing so much as the serpent-hair of the mythical Medusa and her sisters. Grazing sea urchins leave their hiding places at dusk to feed, and brittle stars scuttle across the reef, their sinuous, bristle-covered arms propelling their pentagonal bodies with remarkable speed.

Echinoderms are not the only long-extant animals on modern reefs that seem to have been shunted into the realm of darkness over evolutionary time. Several of the bony reef fishes are themselves practically living fossils. Soldierfishes, squirrelfishes, and cardinalfishes are all members of an ancient order that ruled the reefs while dinosaurs still ruled the land. To ichthyologists, their thick gill covers, heavy, armored scales, and simple, open-and-shut mouths speak eloquently of their ancient lineage. Their eyes are as sensitive as a cat's, but they lack both eyelids and irises to protect against daylight's glare; it is not surprising, therefore, that these fishes prefer to spend their days hovering near the honeycombed walls of coral caves or drifting in the half-light beneath large ledges. When darkness falls, they abandon their resting places and (depending on the species) either hunt for shrimp and crabs along the reef or cruise out into the open water nearby in search of nocturnal plankton.

Rigid coral branches protect tiny blue *Chromis* during the hours of darkness.

Many marine biologists believe that most presently nocturnal animals were originally diurnal. Piecing together the reasons behind their changeover to nighttime activity is not easy because there is no way for us to study the behavior of fossils. We do know, however, that the end of the Cretaceous period—the era of the great dinosaur extinction—saw an explosive increase in the evolution of new fish species in the sea. This great burst of bony-fish evolution produced many of the modern species that now dominate the daytime reef. During their evolution, these newcomers evolved an impressive array of feeding devices and behaviors, presenting the forebears of soldierfishes and squirrelfishes with strong competition in the search for bottom-dwelling prey during the day. Perhaps as a result of this competition, soldierfishes and squirrelfishes began to take up progressively more nocturnal habits, evolving in the process their huge, staring eyes.

Other animals, such as sea urchins, suddenly faced powerful new, diurnal predators—ancestors of triggerfishes, wrasses, and other modern types—that took a heavy toll of echinoderms caught out in the open by daylight. Under this sort of predation pressure, sea urchins that hid by day and moved about only after dark, while the new predators slept, would have had a great survival advantage over those committed to daytime ramblings.

Recent observations of sea urchin behavior on the reefs surrounding the Galápagos Islands lend credence to this explanation of the nocturnal habits of sea urchins in most tropical locales. Large triggerfishes and wrasses are common over most Pacific reefs, and in these places sea urchins such as those of the genus *Eucidaris* remain hidden all day, emerging to forage only after their nemeses have retired for the night. In some isolated spots, including the Galápagos, however, where few large triggerfishes and wrasses live, sea urchins of the very same genus forage out in the open throughout the daylight hours.

It would be both misleading and erroneous, though, to imply that the nighttime reef is merely a last refuge for a collection of drab evolutionary has-beens. The reef after dark has a beauty all its own. Coral colonies, freed from the danger of nibbling by diurnal butterfly and angelfishes, pump their polyps full of water and spread their tentacles up and out in search of passing plankton. All across the reef, sea fans and soft corals, their arms and polyps fully extended, sway gently with the currents.

Other animals out and about at night include the spiderlike arrow crabs, which camouflage themselves with living polyps ripped from the coral colonies over which they scamper. The darkness also conceals the movements of fragile, graceful nudibranchs whose gill plumes float freely in the open water. By day, these delicate, spectacularly beautiful animals would be torn apart by any number of voracious fishes; at night, they roam with relative impunity. They range in size from species less than 5 centimeters long to the 30-centimeter-long scarlet Spanish dancer. Although usually seen gliding along coral rubble in search of food, Spanish dancers also propel themselves through the water at night by passing their flexible bodies through the graceful swirls that give them their common name.

Among the most extraordinary nocturnal fishes, the eerily bioluminescent members of the genus *Photoblepharon* are seen only in the dead of night, and then only in the absence of bright moonlight. This nearly magical species carries a pair of highly evolved light-producing organs beneath its eyes. In these organs, specialized networks of blood vessels encourage the growth of bacteria that continuously emit a cool, blue-green light. The front of the light organ is covered by a retractable lid that can open and shut in a split second. In contrast to other bioluminescent animals and plants—which usually emit just enough light to be seen—*Photoblepharon*'s luminescence is strong enough for the fishes to see by; a single fish can illuminate a small cave brightly enough to see quite well in it. *Photoblepharon* schools produce so much light, in fact, that, along reefs of the politically tense Red Sea, they have more than once been mistaken for night-operating frogmen equipped with diving lights.

Photoblepharon also use their living lights to communicate, by blinking them on and off, and to avoid predators. When chased, either by natural foes or by scientists with nets, *Photoblepharon* takes evasive action, first dashing in one direction with lights on and then blinking off while swimming away on a completely different course. Watching a school of *Photoblepharon* go through these maneuvers is made all the more fascinating by the light organs' uncanny resemblance to pairs of demonic eyes.

Throughout the hours of darkness, these nocturnal wanderers go about their business on the reef. Then, as dawn approaches, they become as nervous and hesitant as their diurnal counterparts do at dusk. Like the demons Moussorgsky might have imagined as he composed *Night on Bald Mountain,* the sea urchins and the nudibranchs, the squirrelfishes and the hermit crabs, seem to curse the light as they glide, scuttle, or swim to shelter in coral caves from the blinding sunlight. Some, such as *Pempheris,* which have searched far and wide for sustenance, return unerringly to the very same caves they abandoned the evening before. Others, such as crinoids, which had selected suitable perches near their daytime resting spots, slowly fold their jointed areas and disappear into the inner recesses of the coral labyrinth. In unison, corals withdraw their vulnerable polyps as deeply as possible into the protection of their limy skeletons and prepare to move their zooxanthellae into the light of the rising sun. By the time the first direct rays of morning penetrate the waters, the creatures of night have disappeared without a trace, to sleep in the shadows through another day.

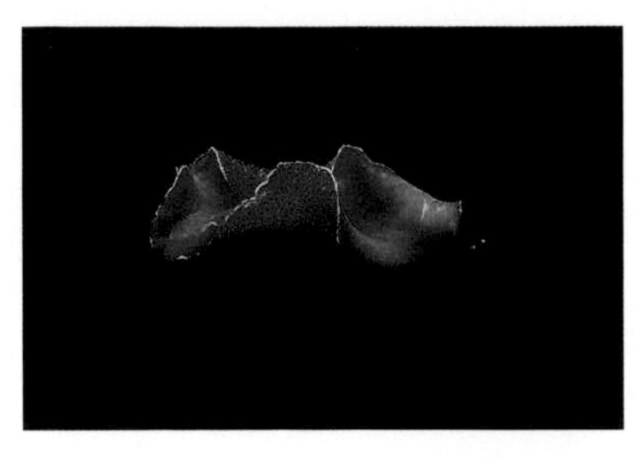

As darkness lifts, a flatworm swims for the shelter of coral rubble.

The Art of Cooperative Living

Following pages:
Whenever it feels threatened, a clownfish dives deeply into the tentacles of its host anemone.

ll over the reef the observant eye spots improbable partnerships, from the cellular level at which corals and their zooxanthellae so intimately depend on each other to the level of large, physically separate animals that are nonetheless inextricably linked to each other by instinctive behavior.

Snuggled deeply amidst the stinging tentacles of a sea anemone—tentacles that would be lethal to most other fishes—the tiny clownfish seems as comfortable as a child in a feather bed. Nestled by the gills of the flamboyant Spanish dancer, a mated pair of shrimp go calmly about their business as their host swims slowly through the water. Hermit crabs tickle the bases of small sea anemones, inducing them to abandon their rocky perches and attach themselves to the roof of the crab's chosen mobile home. Small fishes such as blennies and gobies often share burrows with equally small bottom-dwelling shrimps. The shrimp spends most of its time digging and maintaining the burrow, while the fish uses its sharp eyes to keep a lookout for potential predators. At the first sign of danger, a slight flick of the fish's tail alerts its invertebrate partner and both dive for cover.

It is fitting that the reef—whose very existence is made possible by symbiosis between corals and zooxanthellae—should contain more examples of cooperative living than almost any other habitat. When defined most broadly, symbiosis includes virtually any form of interaction between two organisms. Even parasitism is a form of symbiosis, or "together living"; parasitic symbioses are defined as cases in which one partner benefits to the detriment of the other. Although true parasites rarely kill their hosts outright, they do not make life easier, either.

Another class of symbiosis is commensalism, where one partner benefits and the other appears neither to benefit nor to be harmed. Many reef symbioses fall into this category, from the hitchhiking shrimp on the backs of Spanish dancers to the secretive shrimp and cardinalfish that seek shelter within the defenses of long-spined sea urchins. None of these three beneficiary species is ever found anywhere than with its commensal host, and none ever leaves its host by choice.

Most fascinating are mutualistic symbioses, in which both partners benefit from the association. The most common and tightly knit of these on the reef is the association between corals and their algal helpmates. This sort of extraordinarily close symbiosis occurs not only in corals, but in *Tridacna,* the giant clam, in several species of worms and nudibranchs, and in other cnidarians, including *Cassiopeia,* the "upside-down jellyfish." The reason *Cassiopeia* spends much of its time bottom-up on the floor of shallow bays, in fact, is to expose the zooxanthellae in its tentacles to the sunlight they need. This multitude of tightly knit partnerships enables the reef to support a much larger mass of living tissue than would otherwise be possible in nutrient-poor tropical seas.

Although other reef symbioses are not as tightly connected physically and chemically as these, many are maintained just as powerfully by the behavior of one or the other member of the pair. It is often difficult, however, to determine exactly what benefits accrue to whom. In the clownfish–sea anemone pair, for example, the relationship is a complicated one. Sea anemones, having virtually no central nervous system, can hardly enter into an agreement not to attack the fish. Instead, the clownfish covers itself with a thick layer of mucus to mask the chemical stimuli that help fire the sea anemone's stinging cells.

For years, many biologists thought this relationship was strictly commensal: the clownfish, they believed, obtained protection from its enemies, while the sea anemone was neither helped

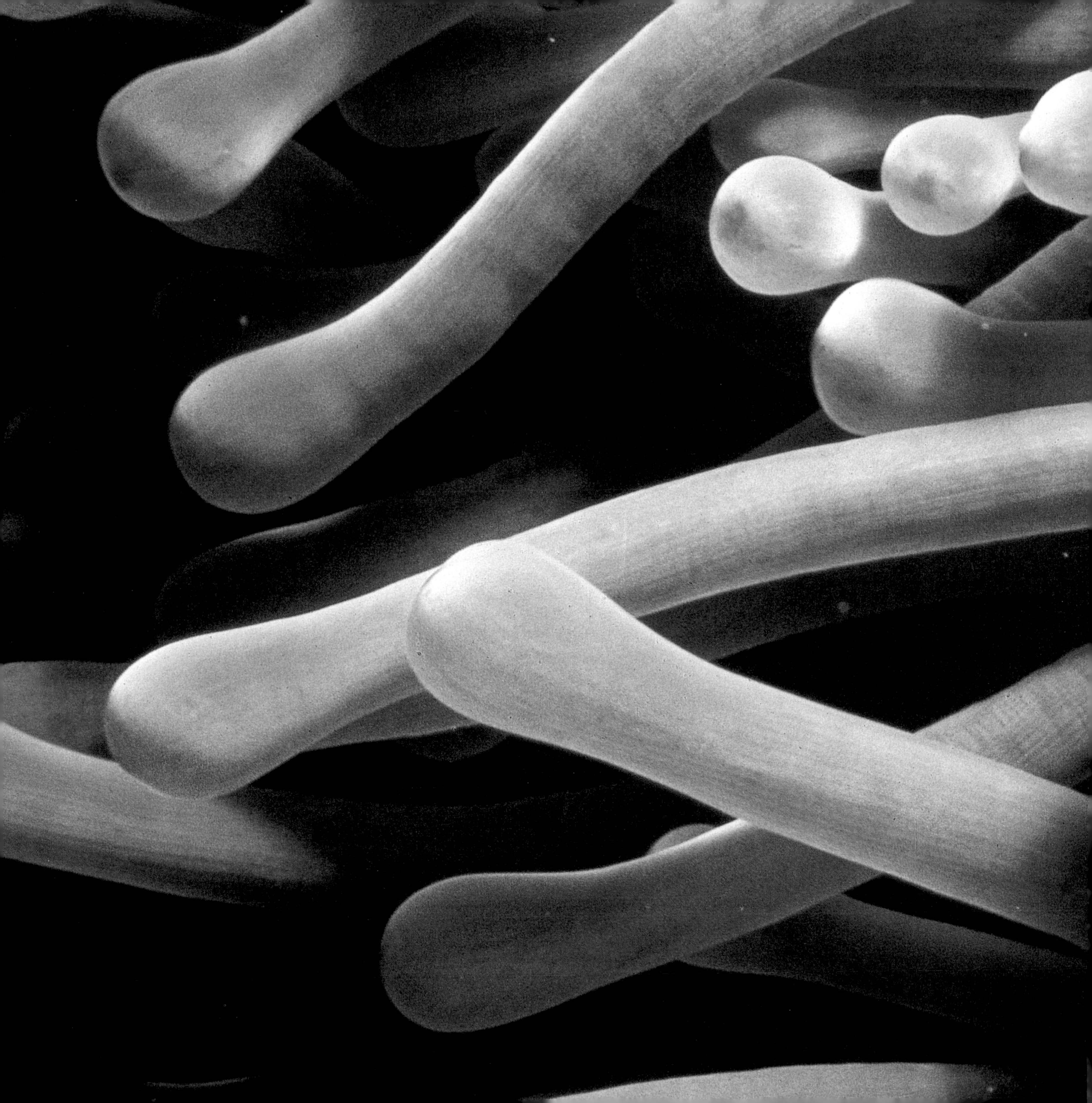

nor harmed. There is still some debate as to whether the fish steals occasional bites of food from its host or brings food items home for the sea anemone to share. There is also some question as to whether the fish's constant mouthing of the tentacles serves some cleaning function for its host. But the relationship is clearly a mutualistic one, as clownfish do protect their sea anemones from predators. After netting a pair of clownfish and removing them some distance from their sea anemone host, biologists observed several butterflyfishes devouring with gusto the anemone's tender tentacle tips. When the clownfish were released, however, they aggressively drove away the marauders, which were many times their own size.

Even experienced biologists are perennially captivated by the behavioral symbiosis of several species of Pacific wrasses that spend their lives removing troublesome parasites from the bodies of other fishes. These brightly banded wrasses form harems that remain within small territories called cleaning stations, whose locations are known both among observant divers and among local reef fishes. Lacking both the appendages of mammals and the flexible necks of birds, reef fishes have virtually no way to groom themselves or to remove the common external parasites that infest them. Instead, they periodically visit their local cleaning station, where their presence triggers a predictable pattern of behavior in the wrasses.

Whenever a potential customer enters their territory, the resident wrasses perform a distinctive zigzag dance to identify themselves as cleaners. Once the visitor identifies the cleaners, it will hover in the water, mouth and gill covers open wide, while the wrasses give it a thorough going-over. Searching the guest's entire body, the wrasses even enter the open mouths and gill cavities of predators such as groupers that could easily swallow them if they chose to do so. As they travel around their customer's body, the cleaners pick off any parasites or loose scales they find, until the host signals the session's end with a shake of its body. At particularly busy cleaning stations, it is even possible to see customers lined up and waiting for their turn.

Superseding the fascination of these symbioses among individual organisms is the ecological importance of symbioses between the reef and the ecosystems it borders. For as clownfish and sea anemones benefit from each other's presence, so the reef, the offshore waters, the shallow-water seagrass beds, and the intertidal mangrove communities nourish and protect each other in ways that scientists have only begun to understand.

A healthy coral reef produces calcium carbonate significantly faster than it is eroded by the surf; reefs in the Caribbean, for example, grow upward at rates between 20 and 40 centimeters each century. By keeping their crests near the surface despite erosion and, in some places, despite the sinking of the rocks on which they rest, reefs form dynamic, living barriers that tame destructive waves. By dissipating wave energy, reefs make it possible for more delicate mangroves and sea grasses to grow in their lee.

Even coral eaters contribute to the growth of nearby mangrove systems. Coral-crunching parrotfish and sea urchins, along with boring molluscs and sponges, turn prodigious quantities of solid coral rock into fine coral sand. Picked up by waves hitting the reef surface, this sand is deposited elsewhere as sediment when the carrying power of the wave-driven currents diminishes. Molded over time by waves and gentle currents, the sediment accumulates as bars and banks in quiet reef lagoons where they can be colonized by mangroves and sea grasses.

Mangroves, in turn, stabilize the sediment further, preventing them from being resuspended in the water by storms or washed out over the reef during heavy rains. Because corals must remain sediment-free in order to function, this protection from recurrent silt baths is crucial to the long-term health of reefs near large islands and major land masses. Corals are also notoriously intolerant of sudden changes in salinity of the sort that heavy rains could easily cause. Many mangrove areas help trap fresh-water runoff from storms, storing it until it evaporates or channeling it through salt beds that often accumulate inland during dry spells. This buffering action helps maintain the stability of the corals' physical environment.

The symbiosis between ecosystems extends well beyond these physical interactions to include the transfer of nutrients and of energy in several forms. Mangrove and sea-grass systems export large quantities of nutrients in the form of dissolved organic matter and detritus, both of which are potential foods for corals and other filter-feeding reef organisms.

Other links among habitats are even more dynamic. Many reef animals actively link the reef with its neighboring systems by making regular migrations between habitats. Members of at least fourteen fish families and several invertebrate groups make feeding forays off the reef. Parrotfish, which sleep on the reef at night, migrate to forage in grass beds by day. Grunts and snappers, in contrast, find shelter among living corals during the daylight hours but slip off to nearby grass beds to hunt after dark. Even painfully slow movers such as sea urchins trundle off the reef after dark to graze in adjacent grassy spots. The significance of these habits to other organisms on the reef is not perhaps intuitively obvious, but as these migrators digest their meals back on the reef, they excrete significant quantities of dissolved ammonia and large amounts of nitrogen- and phosphorus-containing compounds as solid wastes. Recent studies show that the corals among which these animals rest absorb and recycle these nutrients into the mainstream of the reef ecosystem, growing noticeably faster than corals that do not have fish schools associated with them.

In fact, the more one looks for connections between ecosystems, the more one finds. Corals and other sedentary plankton feeders of the reef devour both plankton from the open sea that get washed over the reef by waves and plankton from the grass beds and mangrove swamps that get carried out to the reef by outgoing tides. Recent scientific observation of the Great Barrier Reef indicates that, although corals may well be intolerant of conditions in upwelling zones, they probably do benefit from the periodic blooms of plankton spawned by upwellings in offshore areas. *Pempheris* and other mobile, reef-based planktivores make forays out into open water, hunting plankton that come nowhere near the reef, while reef predators such as barracuda stalk small fishes in their mangrove nursery grounds. Fish-eating birds may eat in one habitat, deposit their nutrients in another, and raise their young in still a third.

Back and forth, day and night, around and around, animals and currents carry energy and nutrients among these systems, mixing their lifeblood and binding their inhabitants together. The lesson here is one of global significance. Mangrove swamps, sea-grass beds, and coral reefs are associations of organisms large and distinctive enough to be separately named by scientists and to be discussed as separate entities. But all of these systems are as interconnected as the tissues and organs of a single animal. Their interdependency stops neither at the shore nor at the edge of the continental shelf: mangroves are intimately tied to terrestrial systems farther inland, while offshore plankton populations respond of necessity to global patterns of air and water currents. Like corals and their algae, like sea anemones and their clownfish, all animals and ecosystems are woven inextricably into the living fabric of the biosphere.

Protected by a behavioral identification code composed of color and movement, a pair of cleaner wrasses removes parasites from the head of a moray eel.

Hatched from eggs laid in the shelter of the
anemone's tentacles, young clownfish become
acclimated to their hosts from an early age.

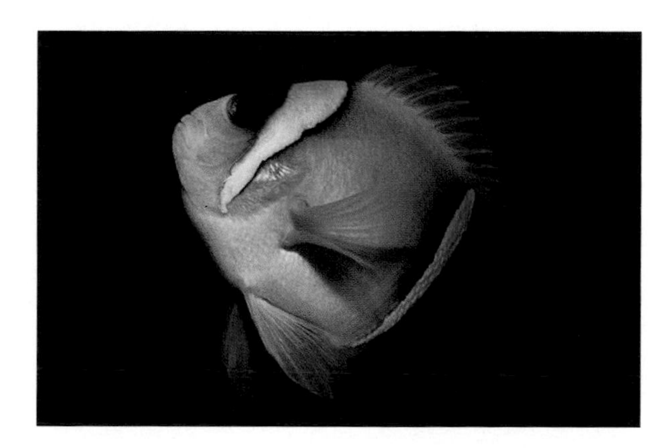

Water Colors

A Living Persian Carpet

Colors abound under the sea, particularly on the coral reef where animals and plants exhibit colors and shapes more varied and bizarre than anything an artist or magician could conjure. Emerald-green algae cling to exposed rock and dead coral. Coralline algae cloak vacant snail shells in gowns of pink chiffon. Animals, sporting angular shapes and bold patterns that could have tumbled from Picasso's *Guernica,* scurry among the coral heads. Fish sweep past in a maelstrom of iridescent fins and scales, creating an interplay of light, shadow, color, and form that could elicit admiration from Monet. And within the reef live creatures sufficiently surreal in aspect to have crawled off a canvas by Dali.

This living kaleidoscope has inspired hyperbole among naturalists since the early eighteenth century, when they wrote that the reef's inhabitants carried "polished scales of gold, encrusting lapis-lazuli, rubies, sapphires, emeralds, and amethysts." To pre-Darwinian biologists like Carl von Linné, the living world in all its diversity was a divine creation; its logic was beyond human comprehension. But the philosophical revolution sparked by evolutionary theory denied to Darwin's successors the convenience of ascribing the damselfish's stripes and the angelfish's mask to the workings of a divine – if somewhat eccentric – artist. Though still awed by reef animals' color patterns, biologists after Darwin have been obliged to explain these phenomena from an evolutionary perspective.

Many early post-Darwinian studies into the function of color in nature were well-intentioned but misguided. Victorian naturalists struggled to identify a single, all-encompassing principle that could explain every case of coloration in nature. Some asserted that all color patterns served to make animals conspicuous. Others, such as the British gentleman-naturalist Abbot Thayer, insisted that every animal's color pattern evolved to camouflage its bearer. Knowing little biochemistry at the time, researchers could not even guess as to whether plant pigments and animal pigments had any direct relationship.

Evolutionary biology has come a long way in recent decades, chalking up both new data and new insights in the areas of animal behavior, ecology, and evolutionary theory. When it comes to modern science's understanding of the functions of color in nature, though, a Victorian scientist might well exclaim in dismay that all this extra knowledge has led to a triumphant loss of clarity; it is now evident that natural phenomena such as coloration are rarely as simple as one might like them to be. There is no single explanation that accounts for the presence and appearance of colors in animals, just as there is no single reason for other phenomena in nature. Rather, natural selection has assembled various color combinations to

A wrasse displays the bold coloration typical of diurnal reef fishes.

Posed on soft corals in a cave at night, a Red Sea starfish demonstrates the sort of contrasting pattern in red and white common among nocturnal and deep-water reef animals.

suit various needs. The colored pigments a species possesses evolve as the species itself evolves: mutations occur in random fashion, providing the inheritable variability that is raw material on which natural selection can operate. Color, structure, and behavior evolve together in the context of the organism's ecological and evolutionary relationships with neighbors and with its physical environment.

Evolutionary adaptations involving vision and color are particularly easy for people to appreciate because, as a species, human beings depend so heavily on sight for sensory information about their surroundings. Human eyes can easily recognize the component parts of visual patterns and can see how the same elements combine to produce different effects. The value of those effects to the organisms that create them is a vivid testimonial to the intimacy that exists in evolutionary relationships among plants, predators, and their prey.

All that Glitters

Both human artists and the living organisms they portray must utilize the physical laws of light to produce their color effects in similar ways. Sunlight, or "white" light, contains electromagnetic energy spanning a broad spectral range from blue (short) to red (long) wavelengths. The presence of all of these wavelengths in roughly equal amounts produces a neutral sensation, so midday sunlight looks white. For an object such as an egret's plume to look white, it must reflect a large percentage of all of the visible wavelengths that strike it. To look black, the fur of the panther—also devoid of what we call color—must absorb most visible light, reflecting very little at any wavelength. Pigments that appear colored, on the other hand, absorb light of certain wavelengths and reflect light of others. Red inks and ripe strawberries reflect a large percentage of the longer wavelengths of light that strike them, while absorbing the middle (green) and shorter (blue) wavelengths.

The fact that colored pigments both absorb and reflect light opens not one, but two sets of evolutionary possibilities for living organisms: either the light absorbed or the light reflected by their pigments can be pressed into service to increase a species' evolutionary fitness. Given the differing needs of plants and animals, it is not surprising that plants have followed one road while animals have taken the other.

As organisms whose existence depends directly on absorbing energy from sunlight, the plants on which most marine food webs ultimately depend have obvious reasons for manufacturing light-absorbing compounds. But chlorophylls alone are often insufficient to meet the energy needs of algae growing in highly colored and light-limited marine environments, and as a result many marine algal groups have evolved accessory pigments to augment chlorophyll's light-catching abilities. Some of these pigments absorb blue light; others trap both blue and green light.

But even as these compounds are absorbing and using the energy in certain wavelengths of light, they are reflecting others. It is the reflected rays—which are useless to the plants themselves—that cause us to perceive different algae as yellow-green, red-brown, or even dark purple.

To the animals that share the waters with this living Persian carpet, this reflected light has enormous significance. To herbivores, reflected light carries critical clues for differentiating among plant species. To animals that must hide, light reflected from algae around them dictates the wavelengths they themselves must reflect if they are to be camouflaged. And to animals that need to broadcast a message over visual channels, the very same colors establish the backgrounds against which those animals must make themselves conspicuous.

Interestingly enough, although the pigments carried by animals often absorb and reflect the same wavelengths of light as the plant pigments around them, light energy absorbed by animal pigments rarely performs any useful work (except in the case of the very special visual pigments in eyes). The energy absorbed when sunlight falls on animal pigments either may be converted into heat or may damage the pigment molecules, causing them to fade. But the presence of these compounds is far from incidental to the animals that carry them. Survival in the sea requires that every bit of energy, every gram of nutrient be utilized to enhance the evolutionary fitness of the species. There is little left over for useless extravagance. If animals' colored pigments offered them no increase in evolutionary fitness, the metabolic energy used to create the pigments could be used more advantageously in some other way; and sooner or later, either through natural selection or through random genetic drift, the pigments would disappear.

In some instances, of course, brightly colored, biologically important compounds are manufactured for purposes having nothing to do with their visual properties. A classic example of such a case was discovered in the inky depths of the Galápagos rift. There, in the total absence of light on the ocean floor, the entire concept of colors is meaningless. Yet searchlights of the deep submersible *Alvin* revealed colonies of giant, tube-dwelling worms totally new to science, each crowned with a plume of scarlet tentacles. Before *Alvin*'s arrival, those worms had never before been illuminated, so this striking pigmentation would seem to be without purpose.

However, the bright red pigment seen by the *Alvin* crew was hemoglobin, the compound that allows red blood cells to carry oxygen. These worms happened to evolve a respiratory system based on hemoglobin instead of on a colorless substitute. Because hemoglobin molecules reflect red light, the thin tube-worm tentacles appear red when illuminated, even though the color has no significance in the life cycle of the worms themselves.

Such situations are not uncommon, but listing all cases in which the purely physiological function of colored compounds is responsible for pigmentation in marine animals would leave the vast majority of animal color patterns unexplained. In most cases, evidence indicates that animals bear special pigments in their bodies only when those colors can be seen, either by other members of their own species or by potential predators and prey. Among most sightless inhabitants of the abyssal ocean floor, where sunlight never penetrates, external coloration is virtually nonexistent. But in brightly illuminated shallow water, where most animals have well-developed eyes, colored animals are everywhere. Coloration, then, must serve to convey ecologically important messages in the form of visual signals. Defining just what messages are carried by animals' colors and patterns of color has occupied naturalists for generations.

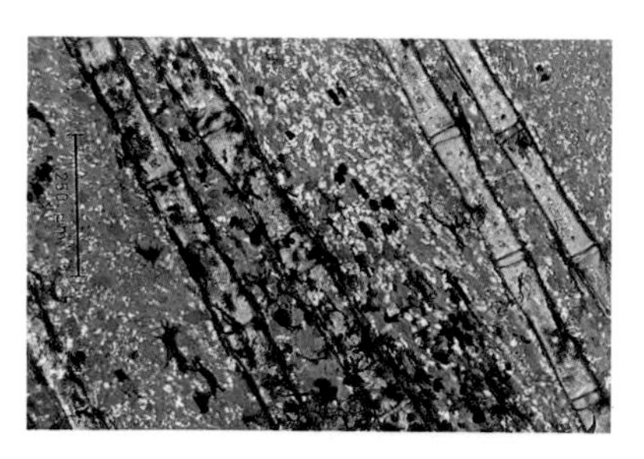

Interference microscopy highlights xanthophores, erythrophores, and melanophores in the skin of a fish tail (© Joseph S. Levine).

The False Front

In piles of grass and algae, on highly colored, highly textured coral reefs, and on the sandy bottoms of oceans the world over lurk animals whose shapes and color patterns have clearly evolved to convey misinformation: that the animals bearing them are not really there. By combining appropriate behaviors and color patterns, many animals manage to hide themselves so well that only experienced eyes could ever distinguish them from rocks or other nonliving parts of the environment. Only when what seems to be a clump of algae suddenly swallows an unwary fish is the charade broken and the animal revealed.

The combination of evolutionary adaptations necessary to effect such masterful disappearing acts is impressive. Not the least of these is the ability to sit still for long periods of time. Acting the part of a rock is not as simple as it might appear; but assuming the shapes, colors, and textures necessary to look like a rock—especially if the model rock is covered with algae of different colors—is extraordinary.

Many marine animals, including chitons, must blend into a multicolored carpet of green, red, purple, and brown algae. Chitons, which are algae eaters, have evolved the ability to sequester ingested pigments from the plants around them into their armadillolike shells. Members of a single chiton species will all exhibit similar patterns on their shells, but the colors in those patterns vary depending on the colors of the algae present in their environments—and hence in their diets. Visually, they are what they eat. This habit of pirating plant pigments is common, for many marine animals are unable to produce certain important colored compounds on their own.

A chiton becomes part of the living tapestry around it by utilizing pigments from the algae on which it feeds.

Some invertebrates camouflage themselves not by means of internal coloration but by attaching bits and pieces of their environment to their bodies. Both tropical and temperate oceans contain crabs called "decorators" that cultivate on their backs a veritable jungle of algae, sponges, and invertebrates of various sorts, enabling them to blend in effectively with their surroundings.

Though these concealment techniques are effective for sedentary animals whose background (from a predator's point of view) remains the same, they create static images not amenable to change and are thus useless for active species that may need to hide against any of several different backgrounds. In adapting to this situation, fishes and certain motile invertebrates have evolved more complicated and dynamic means of generating visual signals—means that involve specialized skin cells, a variety of colored pigments obtained from food, and other pigments the animals themselves manufacture.

Many animals, from squids to human beings, synthesize one or more melanins, compounds that range in color from brown to black. In people, differences in the melanin content of the skin are responsible both for the varied skin color among races and for the phenomenon of tanning; their primary purpose seems to be protection against damaging ultraviolet radiation. Aquatic animals, however, use melanins primarily for visual purposes; and unlike people, many marine species can actively control melanin distribution in their skin.

In such species, melanins are produced in the form of small granules called melanosomes located within large, branching cells called melanophores. In the time it takes a person to blush, the nervous systems of fishes can change the distribution of these granules within the melanophores, causing them either to disperse or to aggregate. In this manner, the animals are able to change their skin shade from dark to light. Fishes change color to signal success or defeat in aggressive or sexual encounters, to improve their concealment after moving to lighter or darker areas, or simply to

register alarm. Anyone who has ever seen aquarium fishes develop, lose, and then regain stripes or bars in a matter of minutes has seen melanophores in action.

When it comes to producing colors other than black, animals have few equals in inventiveness. White, a most useful color to have, is produced by human artists using such exotic chemicals as titanium oxides. Animals accomplish the same result by utilizing one of the most common organic compounds in nature, the amino acid known as guanine. When produced in pure form, guanine crystallizes into tiny, flattened plates that reflect light like microscopic mirrors. In fish skins, specialized cells called iridophores create and arrange a myriad of these crystals in parallel arrays. The ordering of the guanine crystals causes them to reflect light of most wavelengths in all directions, creating a brilliant white surface. By varying the orientation and regularity of these crystals, fishes can also use them to produce the flashing silvery hues so often seen in herrings and minnows.

A layer of guanine crystals can also underlie other pigmented cells in the skin, serving much the same function as the whitewash an artist might apply to a canvas before painting. Bouncing back light that would otherwise be absorbed by the tissues of the animal, the guanine layer adds sparkle and brilliance to other colored pigments spread on top of it.

Still another common visual special effect seen in marine animals is the result of guanine mirrors. Scores of marine fishes, from tiny herring to mighty game fish such as tarpon and sailfish, have iridescent scales over which light rays seem to dance and shimmer in hues of blue and violet. But as any sportsman can attest, this visual show vanishes completely if the scales dry out. Once they are gone, no taxidermist can recreate the colors, and no photographic film can precisely duplicate their appearance.

In most families of fishes, these colors are generated, not by blue pigments, but by a well-ordered array of guanine crystals. When set at just the right angle and spaced at precisely the correct distance apart, guanine platelets create optical interference phenomena in the light rays that strike them. Certain wavelengths pass through the array while others are reflected. The same sorts of interference phenomena generate colors in soap bubbles, where the variety of colors reflected is caused by differences in the thickness of the soap film. The spacing of the guanine crystals in fish skins is regulated to reflect blue light, perhaps as an evolutionary response to the almost total lack of blue pigments in fishes. (There are a few groups of fishes—notably parrotfishes and wrasses—that do manufacture real blue and blue-green pigments, but they are in the minority.)

The interference phenomena behind these colors, in addition to explaining the almost magical interplay of hues in the skins of living fishes, explain their disappearance upon drying. Desiccation of the tissue in which the guanine platelets are embedded causes changes in their spacing and orientation, destroying the interference effect.

Reds, yellows, oranges, and browns are colors human artists commonly make with vegetable dyes. The plants that produce those vegetables usually color them using compounds called carotenoids, which are also quite common as accessory pigments in many marine algae. Many fishes, too, use carotenoids for their visual displays, but they have never evolved the ability to manufacture them. Instead, over countless centuries of random variation, different fish species have evolved ways of assimilating existing pigments that they obtain, as chitons do, by eating plants or other animals that contain them.

Fish funnel these pigments into cells called xanthophores (if the pigment is yellowish) or erythrophores (if the pigment is red). Because of

Preceding pages:
Masters at quick color changes, octopi expand and contract pigment cells to alter their appearance.

A decorator crab obscures its identity with bits of sponge and algae.

the way most carotenoids absorb light, changing the concentration of a carotenoid in a cell—without changing its chemical composition—can actually change the cell's apparent color from yellow through orange to deep red. This characteristic of carotenoids comes in very handy, because it allows a fish to effect significant color changes without major alterations in body chemistry. All a fish need do to change gradually from yellow to orange is pump more carotenoids from food into its pigment cells.

Together, these three chromatophore classes and their pigments enable fishes to generate black, white, blue, yellow, and red. Once an animal evolves melanophores, iridophores, xanthophores, erythrophores, and physiological pathways to fill them with pigments, it can achieve an almost infinite variety of color effects using only simple rearrangements and recombinations of these visual elements. Areas of blue can be generated by combining a guanine layer (to reflect shorter wavelengths) with a thick undercoat of melanin. The melanin underlayer absorbs longer wavelengths of light passing through the iridophores, while the reflected shorter wavelengths produce the color blue. To generate green elsewhere on the same fish's body, the same iridophore layer can be teamed with clusters of yellow xanthophores and a reflecting guanine undercoat. In this configuration, the blue light reflected from the interference filter combines with the yellow light passed by two trips through the xanthophore layer, and the mixture appears green.

This kind of simple flexibility can be critical to the evolutionary success of an organism. Random genetic variation, the source of evolutionary change, cannot be expected to produce very often the sorts of complicated genetic alterations involved in creating new cell and pigment types. Likewise, genetically controlled changes in behavior and digestive physiology are far from trivial, so animals cannot easily effect changes in diet to obtain new colored compounds from different food sources. If the evolution of color patterns in fishes depended on mutations of this complexity alone, such evolution would be hesitant at best. But genetic variations in the arrangement of extant chromatophores or in the concentration of pigments already present do not involve major evolutionary leaps and are in fact quite common in marine populations. Some individuals' melanophores are more or less concentrated than most, and some individuals' xanthophores contain more or less pigment. With this kind of potentially useful variation present in a population, the differentiating pressure of natural selection on the population's coloration can work quickly and effectively, and evolutionary changes in color and pattern can occur in short order if the need arises.

The number of different effects that can be produced by rearranging and recombining the few basic color elements possessed by marine animals is staggering, and luckily so, for matching a background closely enough to blend into it requires both precision and a varied palette to choose from. As any artist knows, lightness is as crucial as hue in effecting a proper color match, and changing the lightness of a pigment can make the mixture change in hue as well. Lemon yellow, when darkened with black, becomes lime green; red and orange, darkened in a like manner, change to shades of brown. Fishes are able to make use of these visual effects on their bodies: by expanding and contracting their melanophores, they can create the same effect as if they increased or decreased the amount of black in their pigment mixture. Using this simple physical procedure, fishes are able to generate a phenomenal range of shades and tints without altering their carotenoids in any way.

Because melanophores can expand and contract within minutes, fishes and other animals that have evolved the ability to control these pigment cells can change their appearance very quickly. So effective are the camouflage techniques of flatfishes that some observers credit them with the power to change both their colors and their patterns in a matter of minutes. Actually, they are able to change neither their colors nor their body patterns rapidly enough to hide themselves. But color and pattern are not the only factors that make an object visible against its background; brightness, contrast, and sharp edges are also important visual cues, and flatfish have evolved the ability to alter these aspects of their appearance by behavioral and chromatophoric manipulation.

Flatfish can match the general level of background brightness and contrast with remarkable accuracy by manipulating melanophores. This method works as well as it does only because flatfish simultaneously blur the visual boundary between their bodies and the region immediately surrounding them. Scurrying away from an intruder, a flatfish races for the ocean floor in a flurry of fins and tail, thereby stirring up a cloud of sand as it reaches the bottom. When the sediment finally settles, much of it falls haphazardly on the fish's fins, obscuring the sharp edges so critical to visual perception. This behavior reflects an apparently innate understanding of their limitations. These fishes never try to hide on corals or sponges, and they rarely rest for long on pebbles or rocks. They avoid all such areas in favor of the more neutral and less strongly featured sandy bottoms they are so well-adapted to imitate.

The peculiar goosefish and the incredibly well-concealed scorpionfish, however, are far from defenseless creatures that rely on camouflage for protection. Concealment can be used in offensive strategies, too, and quite a few predators make use of camouflage to hide themselves from prey.

Goosefish, scorpionfish, lizardfish, and stonefish are but a few of the ocean's ambush predators—sedentary carnivores who lie in wait for prey unwary enough to swim within range of their jaws. Since many potential food species have excellent eyesight, ambush predators have evolved body pigment patterns and textured skins that enable them to blend perfectly into their favorite hunting environment, be it a sand flat or a multicolored backdrop of corals or sponges.

Such background matching is far from easy; on a coral reef, at least, it is rather like a chameleon trying to hide in an Oriental rug shop. Each coral head and sponge-encrusted rock has its own combination of textures and colors. Pebbles and stones vary in size, hue, and lightness. An individual predator could never perfectly imitate the visual characteristics of all possible backgrounds in a given habitat, so each individual spends most of its time in a restricted territory. There, over time, it duplicates the appearance of the elements present both by melanophore/iridophore manipulations and by long-term control of the carotenoid content of its xanthophores. Even so, the camouflage works only if the fish remains motionless in front of the appropriate background; if it moves or is displaced, it may stand out in stark contrast to its surroundings.

Despite fishes' mastery of disguise, they are no match for the octopus in speed and adaptability. Most of the octopus's cephalopod relatives, including common squids and cuttlefish, have a great deal of control over their chromatophores, but none can flush and fade or blanch and darken as swiftly as an octopus. When excited, octopi can send wave after wave of color rippling over their bodies; when they want to hide, they can melt into their backgrounds in an instant. Like flatfishes, they direct their short-term changes in coloration toward matching the brightness and contrast of their backgrounds, but like scorpionfish they seem able over longer periods of time to match with their pigments the specific colors of their environment.

Surprised at his burrow entrance, an octopus begins to match the color of nearby rocks in seconds.

On guard at the entrance to its tunnel, a fringehead fish matches not only the color, but the shape and texture of the surrounding algae.

Fringed appendages and a mottled surface make goosefish tough to spot on muddy bottoms off New England.

Extraordinary accuracy in color and texture matching allow a scorpionfish (*above*) to disappear in a bed of seagrass and algae; a related species (*below*) sits, nearly invisible, on silt and algae-covered rocks.

The Public Disclosure

Yet another motionless scorpionfish imitates both the color and texture of the coral outcropping on which it rests.

The concealing coloration of camouflaged animals is an integral part of a total life-history strategy that requires them to remain motionless and hide, either to avoid predators or to catch their own food. But many coral reef animals have evolved along totally different ecological lines, with feeding requirements or social behaviors that simply are not served by such a sedentary style of life. Butterflyfishes swim briskly around the reef all day, searching for specific coral polyps, algae, or shrimp that make up their specialized diets. Many wrasse species congregate in large schools that mill around continually, maintaining a dominance hierarchy, feeding, and mating. Many of these species, which act so differently from the camouflage crowd, sport electrifying visual displays of bright colors and bold patterns that defy simple explanation.

Some biologists have argued that, regardless of how conspicuous these color patterns look to the naive observer, the patterns actually function as aids to concealment in the natural environment. According to this thesis, bold, irregular patches of light, dark, and color break up the familiar features of an animal's body and distract attention from its outline. If those patches happen to match objects in the animal's background in color and shape, so much the better; but visual disruption, not background matching, is the critical factor. The classic examples that support this hypothesis are the zebras of African savannas and the sargassumfish of tropical oceans. Both of these animals, ordinarily quite mobile, combine disruptive coloration with a tendency to "freeze" when frightened. Their ability to stand still when threatened works together with their body colors to allow the animals to blend into the visual patterns created by the tall grasses of the savanna or the tangle of sargassum weed.

But this hypothesis cannot explain the stripes and colors of the many damselfishes, butterflyfishes, and angelfishes, none of which "freezes" when frightened. Depending on the circumstances, these species either flee across the reef or converge into a school whose members dart quickly back and forth. These fishes cavort across the reef in broad daylight bearing colors that fairly leap out at observers, and their patterns often contain elements that precisely delineate, rather than conceal, their body outlines. Color, pattern, and behavior in such cases clearly draw attention to their bearers, rather than concealing them. The assertion of conspicuousness does not mean that these color patterns cannot serve to protect their bearers from predators, but it does indicate that the protective mechanism is not camouflage in the traditional sense.

The major difficulties in explaining conspicuous coloration are twofold. First, an ability to hide from predators does make sense as a survival tool. The reef, with its population of barracuda, sharks, and jacks, does not lack voracious fish eaters. Even in environments where predators seem much less common, the evolutionary advantage of camouflage has dominated the evolution of body coloration, making conspicuous coloration relatively rare. Yet hundreds of reef dwellers seem oblivious to this phenomenon. Second, conspicuous coloration does not confer any intuitively obvious benefit that might sustain certain species along evolutionary paths where the benefits of concealment are discarded.

Understanding why reef animals in particular have become so brightly colored requires information from several fields of inquiry, for the interconnectedness and complexity of life on the reef precludes a simple answer. To begin with, color camouflage is critical for a species only if its most important predators use vision to locate prey during the daytime, when sufficient sunlight is present to allow color vision to function. Recent research has shown that this is often not the case on the reef. The most important vision-dependent reef predators—aside from the camouflaged ambush hunters—either do not hunt during the day at all or rarely succeed in catching anything when they do hunt during sunlit hours.

The reason for this is that the keen eyes of diurnal reef fishes, adapted strictly for use in good light, provide their bearers with sharp vision that is well-equipped to distinguish colors and movement. During the day, therefore, sneaking up unnoticed on these diurnal species is no easy task. Furthermore, many of these same brightly colored reef species are agile swimmers that hang close by the reef itself. As soon as they spot potential enemies, they dive with lightning speed into narrow, twisting reef tunnels. No barracuda or jack can wend its way through those tortuous passageways, and in fact large predators rarely even try to hunt near the reef by day.

But while the sharp eyes of diurnal fishes give them a competitive edge during daylight, they do not function as well during the visually confusing dawn and dusk twilight periods. Fish-eating predators, in contrast, use their large, sensitive eyes to stalk during the semidarkness of twilight, while their prey is more vulnerable.

Adaptations in nature involve trade-offs, however. Like old-fashioned, high-speed photographic film, the eyes of these predators provide grainy visual images; they also are not particularly good at distinguishing colors. The twilight predators as a group can neither see colors nor discern finely detailed images.

Strictly diurnal species thus have as their principal enemies either ambush predators, which depend for prey on creatures of any coloration blundering too close, or twilight predators, which are basically color-blind. Under these circumstances, the presence of bright colors and intricate patterns on the diurnal fishes' bodies does not greatly increase their risk of being eaten. For this reason, if diurnal species have no need for cryptic coloration to help them obtain their own food, they experience no evolutionary pressure to conceal themselves visually. If, then, any evolutionary advantage were attainable through conspicuous coloration—no matter how marginal that advantage might be—selection for bright colors would be unopposed by contravening pressure for protective coloration.

Konrad Lorenz, the Nobel Prize–winning Austrian ethologist, first proposed the sorts of advantages conspicuousness might bestow. From his studies in animal behavior, Lorenz knew that animals often use prominent visual markers to identify themselves and to advertise their sex and physical condition. When he first observed a living reef community, Lorenz was deeply impressed by the striking color patterns he saw there. He even invented a special name for color patterns of reef fish: *plakatfarben,* or "poster colors" —a term he used to emphasize his conviction that the purpose of the color patterns was to turn these animals into living billboards, advertising important information.

The reef, Lorenz reasoned, is a community of highly specialized, resident feeders. If the amount of specific food items present in the reef and the number of territories in which those food items can be found limit the population size of each species, it is to an individual's advantage to establish and defend a home territory against competitors. Territorial defense requires a substantial expenditure of energy, however, and on the densely populated reef it would be energetically costly and wasteful for a territorial individual to attack an endless parade of strangers that might or might not actually threaten its particular food supply.

Lorenz reasoned that any behavioral and physical characteristics that helped distinguish genuine competitors from the much larger number of non-

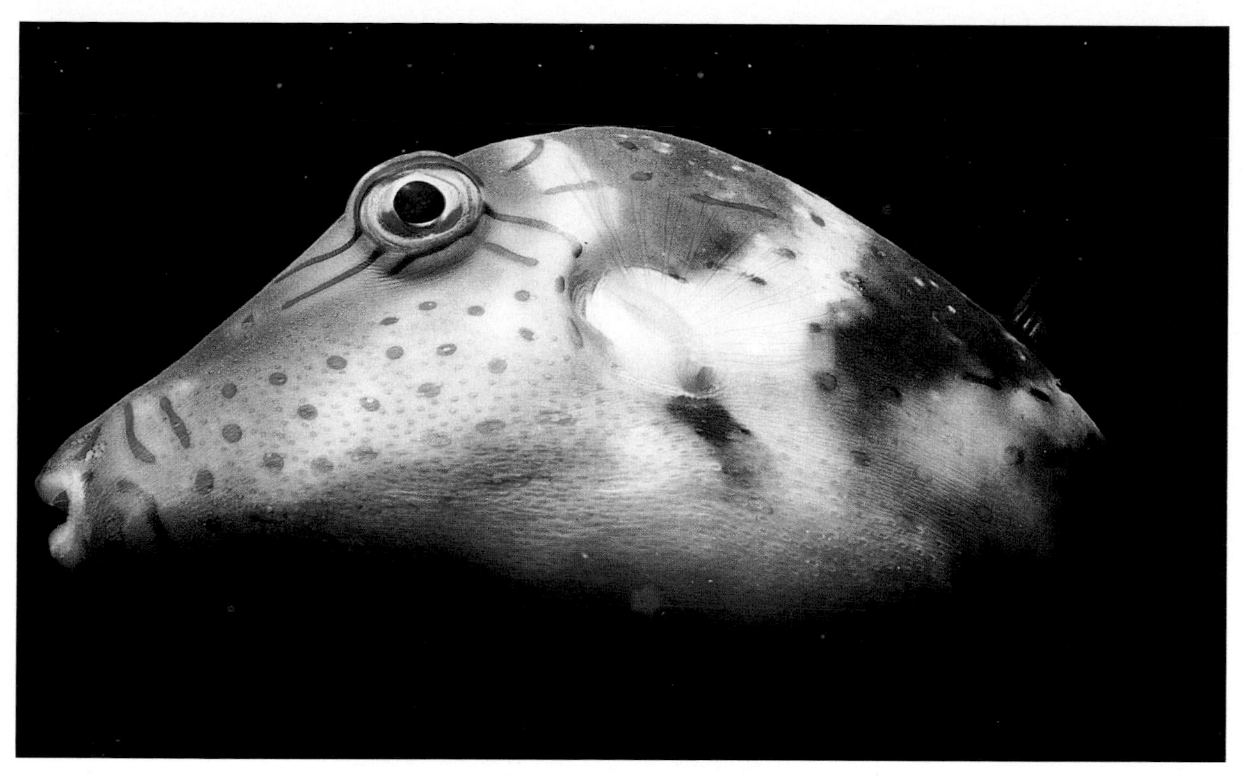

This boxfish is but one reef species whose color patterns serve purposes as yet little understood.

This tropical nudibranch uses its conspicuous coloration to warn potential predators of poisons it has accumulated from its food.

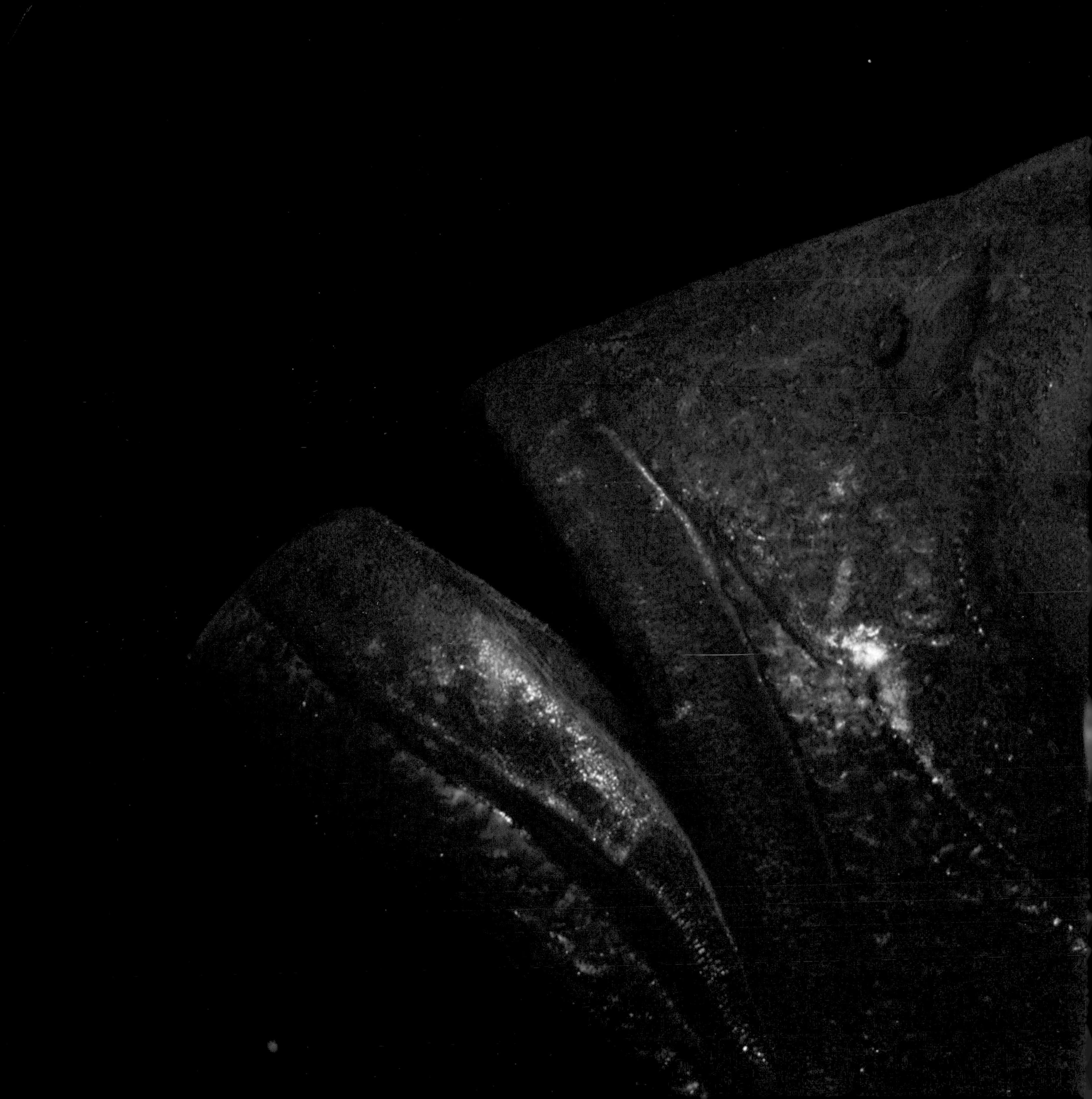

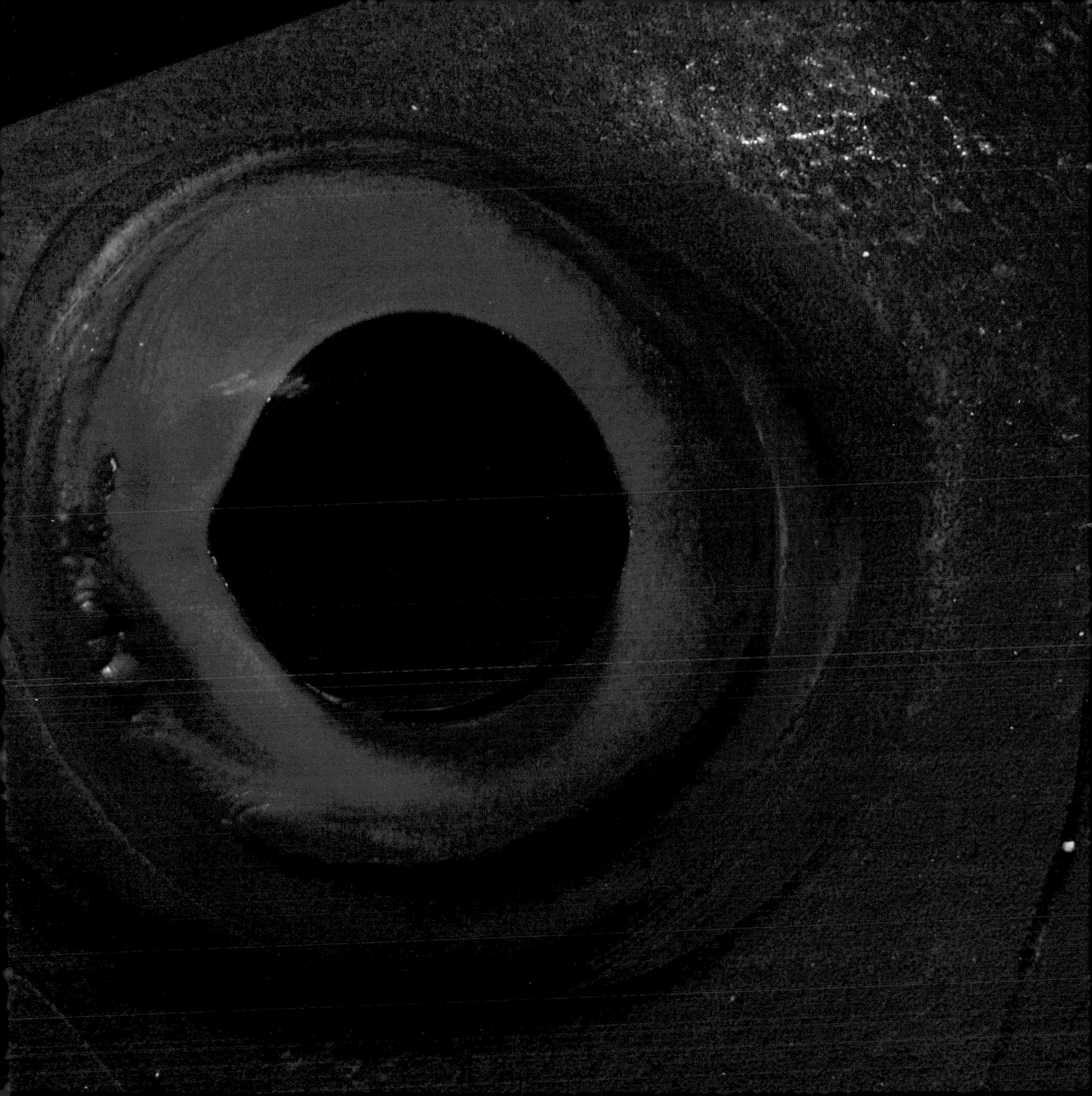

competitors would eliminate unnecessary and counterproductive warfare. Likewise, any mechanism that helped space out competitors by noncombative means would minimize debilitating battles among those individuals. Highly conspicuous patterns on a resident individual in its territory can serve both these purposes by informing other fishes from a distance with a visual message that is as clear as the territorial calls of birds or monkeys.

Lorenz's hypothesis was attractive. It was refreshingly simple and straightforward, it incorporated what was known at the time about the behavioral ecology of reef fishes, and it fit in well with the popular concept of evolution as a goal-oriented process. Lorenz's term "poster coloration" is firmly entrenched in scientific literature, indicating general agreement with the concept that the color patterns of many reef fishes serve as communication mechanisms.

But a great many behavioral researchers have worked on coral reefs since Lorenz made his observations, and data they have collected show that the situation is substantially more complicated than Lorenz believed. Several fiercely territorial damselfishes are not poster-colored, and numerous flamboyantly hued butterflyfishes are neither territorial nor unusually aggressive toward members of their own species. As a result of these and other observations, many biologists today favor modified versions of Lorenz's original hypothesis. The new hypotheses retain the concept of poster coloration as a source of social signals among members of a species, but they recognize that such patterns can serve purposes other than territorial defense.

For example, *plakatfarben* can provide reliable cues to the identity of prospective mates. Diurnal reef fishes with acute color vision can easily use visual signals generated by body color patterns to advertise their gender and their readiness to mate. The more flamboyant the color pattern is, the more readily the message can be read from a distance. The less like any other the pattern is, the smaller the possibility becomes of an accidental mating with a member of the wrong species. Then, too, if highly territorial males are likely to attack other members of their own species, females with species-specific and gender-specific color patterns could gain safe access to males' territories for breeding.

This modification of the Lorenz hypothesis can explain some of the diversity in color patterns among reef fishes; male *Anthias* are distinctly different in appearance from their harem members, and some wrasse and parrotfish species have males and females so differently colored that they were originally thought to belong to different species. A single species of parrotfish or wrasse may, in fact, have four or more different color patterns— one each for immature males, immature females, mature males, and mature females. Presumably, prospective mates are less confused than biologists by this surfeit of visual messages.

Yet here, too, a universal rule proves elusive, for both sexes of many conspicuously colored reef species such as butterflyfishes carry virtually identical color patterns. So nearly identical are males and females of many species that researchers can sex them only by dissecting them and looking at their gonads, by observing them in the act of spawning, or by detecting sex-related differences in behavior. This last method is presumably the one used by potential mates.

A conspicuous color pattern is not necessarily devoid of protective value, but the protection it offers is of an entirely different sort from that offered by camouflage. If conspicuously colored individuals live in large groups and withdraw into tight schools when threatened by a predator, the visual impression they present is one of constant change and confusion. Each one of the individual, highly visible fish in the school darts back and

Preceding pages:
This soldierfish, like many other ancient, nocturnal species, carries vibrant red pigments of unknown function.

These fishes sport the stripes, bars, and spots typical of poster-colored reef species. For schooling sargeant majors (*top*), the pattern probably confuses predators; for clownfish (*center*), it may serve in intraspecific communication; for butterflyfish (*bottom*), it conceals the vulnerable eyes.

forth, commanding visual attention before disappearing into the center of the group. This behavior continually distracts the attention of a predator from one individual to another in the school, making it difficult for the predator to select and zero in on any single fish for the final attack.

Conspicuous patterns can also offer protection by confusing predators as to which end is up—or, more accurately, which end is front. Fish after poster-colored fish sports a conspicuous bar or stripe traversing the head and passing uninterrupted through the eye. In many cases this marking makes the eye itself difficult to spot, even for a human observer who knows in advance where to look. This eye camouflage is often combined with some kind of "false eyespot" elsewhere on the body of the fish, which may lead a predator to expect the fish to dart one way when actually it is preparing to escape in the opposite direction.

Another possible function of conspicuous colors and patterns was proposed by no less a figure in evolutionary theory than A. R. Wallace, who saw many brightly colored tropical animals while working as a naturalist in the East Indies. Wallace observed that many conspicuously colored insects, reptiles, and amphibians were either highly poisonous or extremely distasteful to animals that made the mistake of eating them. If a predator took a bite out of one of these species, it encountered a nasty taste, a venomous counterattack, or subsequent serious illness. Henceforth, that predator would associate the bright color pattern with gastronomic distress, and, for as long as the memory lasted, would avoid eating other individuals bearing similar patterns. The more conspicuous the color pattern or—as described by the Victorian naturalist E. B. Poulton—the more "public" the advertisement of unpleasant attributes, the more likely predators would be to remember the experience, and the more effectively the potential prey would be able to avoid experimental "tasting."

Wallace called markings that function in this way warning coloration. Researchers working since Wallace's time have shown repeatedly that warning coloration has been a powerful factor in shaping the evolution of color patterns in both temperate and tropical butterflies and in a number of other insects. Some researchers have proposed a similar warning function for the striking colors of certain marine nudibranchs, several species of which are either poisonous or venomous. Other reef biologists suggested long ago that predators could learn to associate the bright colors of reef fishes with sharp spines or disagreeable taste.

In all of these cases, poster coloration is most likely to maximize evolutionary fitness when it maximizes the legibility of the advertisement it carries. Consequently, optimization theory predicts that natural selection will favor colors and patterns that make those advertisements as clear, as legible, and therefore as powerful as possible in the reef environment. This seems to be the case; in examining the elements of poster coloration, biologists have found the same sort of perfection in visual signal function that is found in camouflage coloration.

If it is to stand out as strongly as possible from its surroundings, a single-colored area on a fish's body must contrast with adjacent areas in brightness, in color, or in a combination of both. The components of a complex pattern, in order to be conspicuous, must differ both from each other and from the background in the same ways. As human artists and designers are well aware, color is not at all necessary to make a pattern conspicuous; black and white stripes are highly visible, regardless of the color and intensity of light hitting them. Patterns created with colored pigments offer more diversity than simple black and white, but they need more light in order to show up well, and they may differ in visibility depending on the color of the light in which they are seen.

Reef fishes display their color patterns in soft, turquoise light that both illuminates them and provides a background against which they must be seen. Their reputation for flamboyant coloration notwithstanding, many have evolved patterns of attention-getting stripes and spots of black and white. These patterns are generated by the alternating of areas of quanine crystals with regions containing dense groupings of melanophores.

But perhaps because of the need for many unique, species-specific patterns, reef fishes have often chosen to add color contrast to their advertisements. When poster-colored species do display color, the contrasting color elements are so well chosen that they seem to leap out at observing eyes. Such colors must differ from each other in the wavelengths they reflect and must also differ distinctly from the background light. Yellow and deep violet, the colors most common among reef fishes, fit these requirements perfectly and are simple for fishes to generate using their standard chromatophores.

Strictly diurnal, shallow-water species such as parrotfishes and wrasses spend most of their time in areas where light is bright and contains a broad range of wavelengths. The range of effective colors broadens along with the level of illumination, and these species regularly include greens and oranges among their colors.

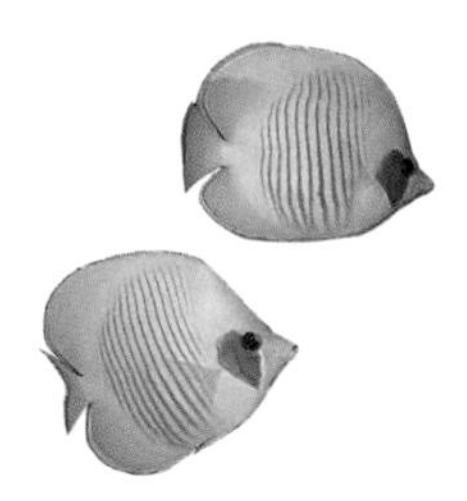

Changing Colors for Changing Needs

Poster-colored reef fishes are a special case; they have the reef to hide in, have few diurnal predators, are quick and agile, and have ample ecological justification for making themselves conspicuous. On the other hand, species such as flounders, scorpionfish, and anglerfish have experienced consistent evolutionary pressure from predators to blend in with their surroundings. But for most organisms, the evolutionary choice between concealment and conspicuousness has not been a simple one. Any animal threatened by visual hunters can best avoid being eaten if it makes itself as inconspicuous as possible. Visual advertising for a variety of purposes, however, is best conveyed through conspicuous coloration.

How this difficulty is resolved depends on the versatility of the various fishes' body-coloring systems and their ability to change color. By adopting camouflage colors for most of the year and switching to conspicuous colors exclusively for their mating season, many species benefit from the protection of the former and at least one major advantage of the latter. Well-camouflaged for most of the year, Atlantic sea ravens become intensely colored during their breeding season. The precise colors they assume vary, depending on where they live, from yellow—the most visible color in blue water—to the oranges and reds that are more clearly seen in the green waters of the temperate North Atlantic. Many fresh-water tropical fish species also exhibit this physiological adaptation, and fresh-water aquarium hobbyists are familiar with the varied and brilliant colors acquired during breeding activity by many aquarium fishes.

Animal coloration provides convincing testimony of the creativity of natural selection. For all its beauty and complexity, however, the relationship between color and behavioral ecology in marine organisms is only a fraction of the wonder to be found in contemplation of their evolution as a whole. The entire realm of color phenomena—with all the complex interconnections of underwater light, photosynthesis, the visual abilities of predators and prey, body colors, feeding and reproductive behavior, and ecology—is but a single dimension in the sensory existence of these organisms. Every phenomenon involving vision and color is matched by a corresponding phenomenon involving smell and odor, taste and flavor, hearing and sound, and other realms of sensory perception entirely outside human experience. Each sensory channel is controlled by its own intricate genetic mechanism, and each must function within each particular organism's physiological and ecological environment. From a broader perspective, each of these avenues of information is essential in linking organisms together in the complex living web of ecological interactions that forms a biological system.

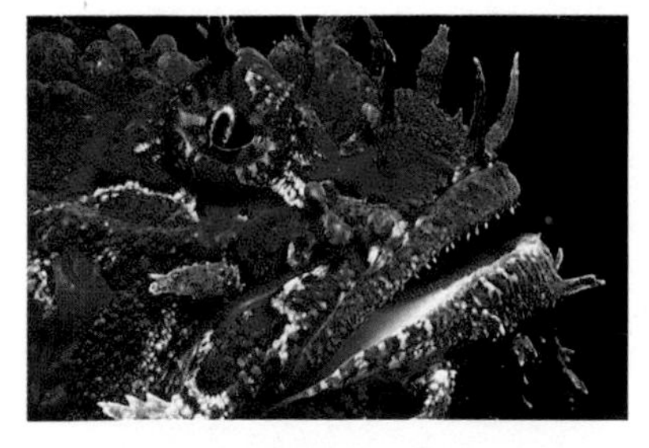

Sea ravens, normally drab in hue, use color to advertise their availability during the breeding season.

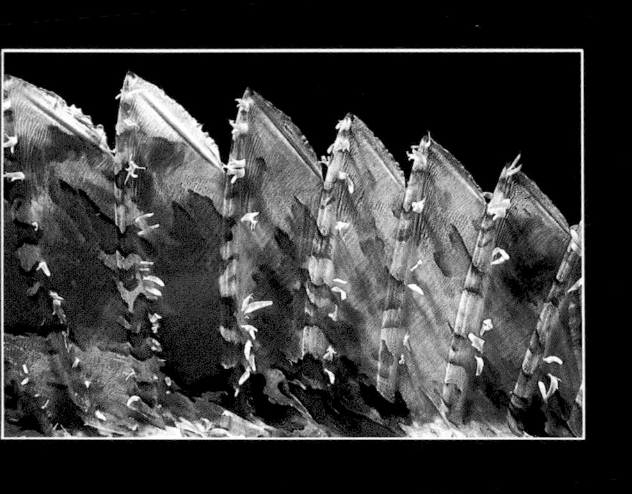

Poison, Tooth, and Claw

The Undersea Arsenal

Dorsal spines of scorpionfish carry venoms that damage both nerve and blood cells.

Following pages:
This New England jellyfish, like other *Cnidarians,* carries envenomed stinging cells on its tentacles.

Disturbed during its nocturnal wanderings, a puffer pumps itself full of water, transforming its normally slender body into a grapefruit-sized gargoyle. Nearby, an imperturbable lionfish glides through the water, rippling its long, venomous spines in the current like the veils of a latter-day Salome. Half a world away, a long-spined sea urchin in search of its nightly meal of algae bristles with brightly colored barbs, each of them waving about individually as its owner plows ponderously along the bottom.

As long as people have known the sea, encounters with such macabre and dangerous creatures have aroused fear of both real and imagined ocean dwellers. Mermaids reportedly lured men to watery graves, giant clams trapped unwary swimmers, and death-dealing demons lurked in every submarine cave. Modern undersea explorers have explained away most mythical fears: mermaids were probably manatees seen from a distance, giant clams trap only those foolish (and forceful) enough to wedge a limb between their shells, and cave-dwelling sea monsters were not dragons but moray eels. Today, the weightless, multicolored world of the coral reef and the verdant, fluid realm of the kelp forest retain Neptune's majesty while shedding his menacing trident. Such confidence is generally well-founded; shark phobia to the contrary, human beings are preferred prey of very few undersea creatures.

Marine animals themselves, however, face a constant battle for life and livelihood in an environment as perilous as its vistas are scenic. Fishes and invertebrates alike must locate and capture prey; simultaneously, they must avoid becoming the prey of others. Sedentary organisms on coral reefs, faced with intense competition for living space and surrounded by bacteria at every turn, must fight off infections and must prevent spores and larvae of other organisms from settling on them.

Under siege in this manner for eons, reef animals have evolved an array of poisons, weapons, and defensive behaviors as bizarre and as potent as any imagined by Jules Verne. Marine biologists study these animals for their evolutionary and ecological interest, and medical researchers probe the chemical and physical arsenals natural selection has provided them in search of medically useful compounds. Venoms that incapacitate the nervous system provide tools for studying nerve cell function. Chemicals evolved in the course of millennia of struggle for survival among marine invertebrates hold promise for scores of useful antibacterial, antiviral, and even anticancer drugs. And protective behaviors offer insights into behavioral ecology and into the evolution of behavior.

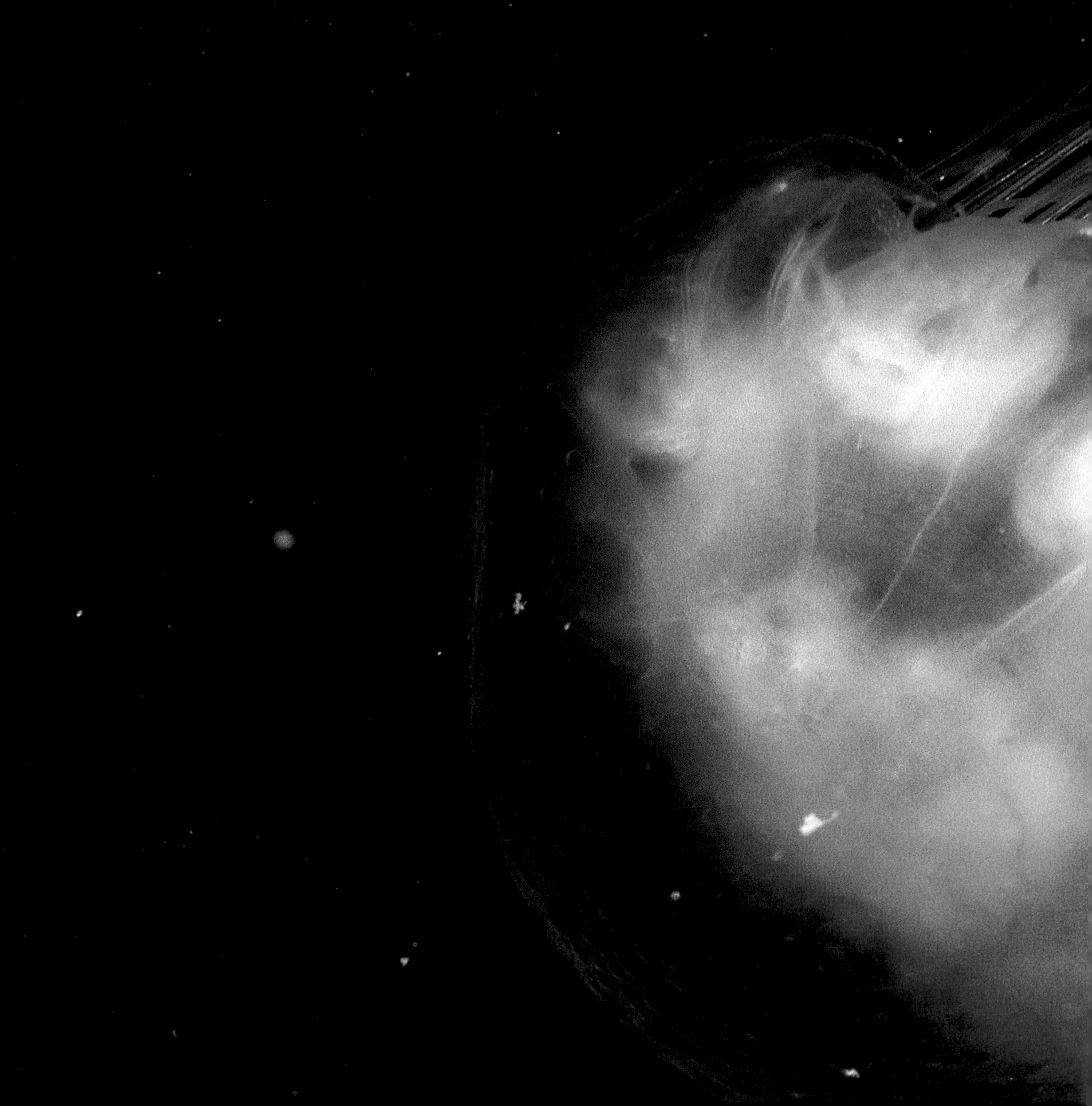

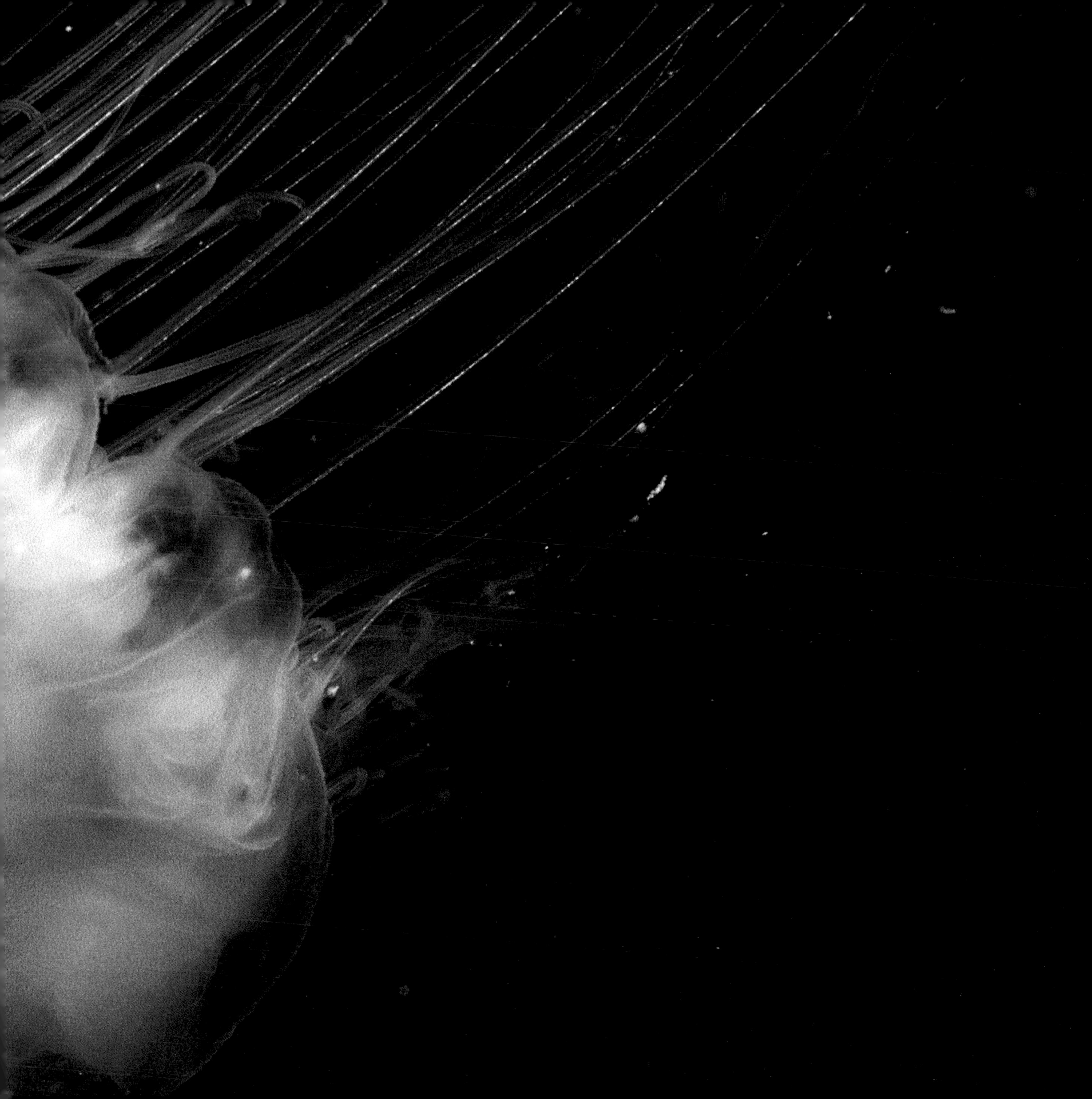

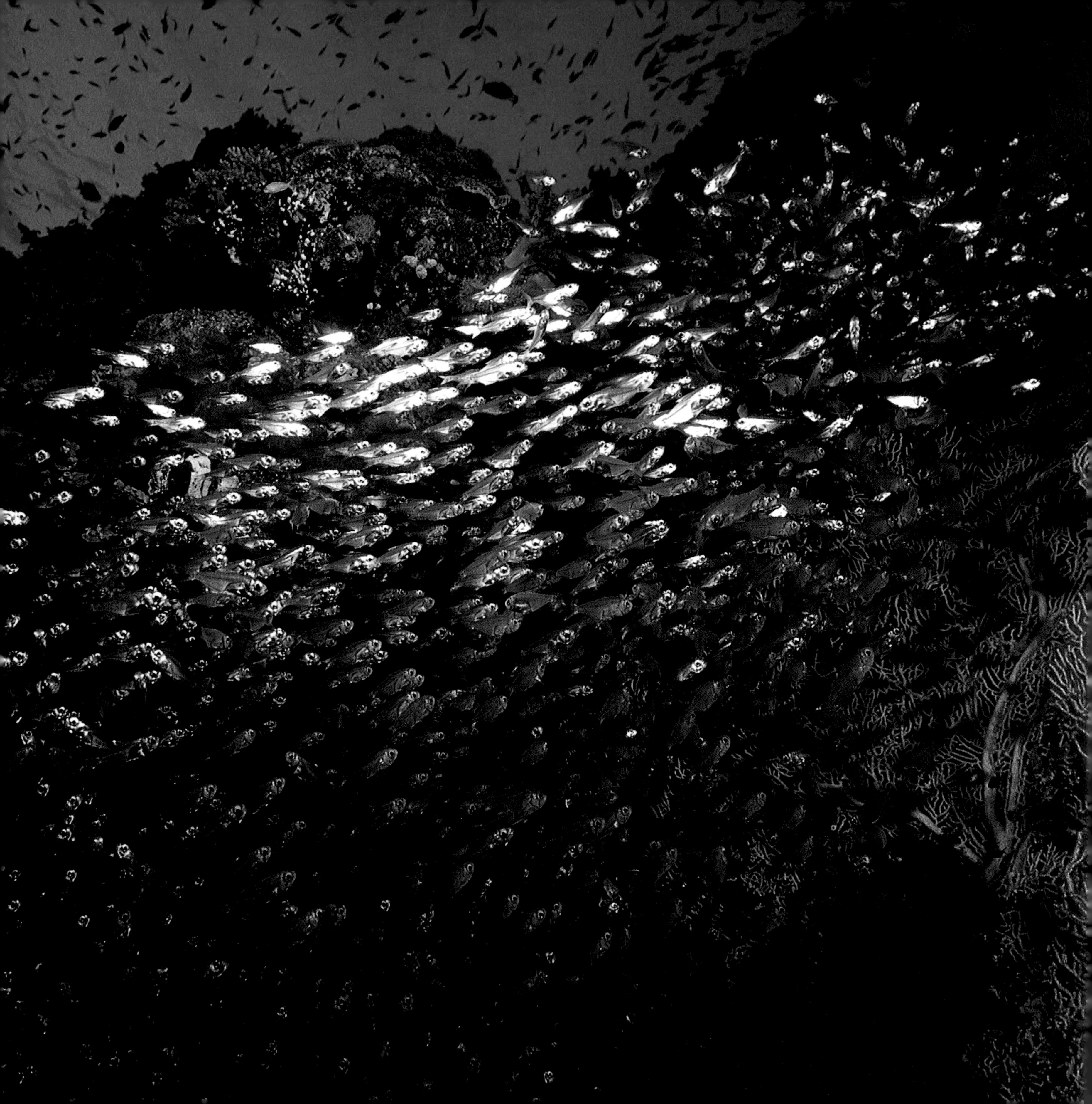

Strength in Numbers

Many characteristics, particularly behavioral ones, of marine animals have clear adaptive significance. Innate behaviors that may at first have been casual and loosely formed have become ingrained over evolutionary time into tightly controlled, stereotyped actions such as schooling that protect against predation, afford the animals a competitive advantage over other species, or facilitate feeding. Milling around in a flowing, kaleidoscopically shifting swarm, small fishes such as damselfishes, herring, *Anthias,* and *Pempheris* find safety in numbers; large schools of moving individuals fill their predator's visual field, making it difficult for the predator to focus an attack on a single target.

Browsers such as goatfishes use schooling as an effective way to search featureless sand beds for food; a group can spread out to investigate a wide swath of sand, converging on local concentrations of worms and other buried prey turned up by individuals in the school. Algae-eating surgeonfish gain safety and feeding ease in large, milling throngs that travel long distances in search of green pastures. When they happen upon a territorial species whose aggressive defense usually keeps other fish at bay, the surgeonfish descend en masse to overpower the hapless resident and denude the area of all edible plant material.

Poisons and Venoms

Equally important in both defense and feeding is the wide variety of toxic compounds manufactured by marine life ranging from algae to fishes. Many toxic compounds, by-products of metabolism that were at first coincidentally harmful to organisms other than those generating them, have been refined through natural selection to the point where they now protect their bearers from bacterial or viral infections, discourage other organisms from settling on top of them, make them unpalatable or dangerous to eat, or enable them to overpower their prey. Other biological poisons, whose potency is no less formidable, apparently have no "useful" function; these present unsolved mysteries to biologists still convinced that every characteristic of every organism evolved to serve some critical biological purpose.

Toxic chemicals in nature fall into two basic groups, which are highlighted by the title of Dr. Bruce Halstead's magnum opus, *Poisonous and Venomous Marine Animals.* The term poison, although it can be used for all toxic compounds, refers in scientific parlance strictly to substances that must be ingested in order to affect an organism. The term venom, on the other hand, is used for toxins that are physically injected into victims by spines, fangs, or stinging cells.

Practical knowledge of specific poisonous fishes dates back to ancient written records from the Middle East and China, although that knowledge has often been shrouded in mysticism, speculation, and exaggeration. Poisonous puffers have been identified among fifth-dynasty Egyptian hieroglyphics, and certain biblical dietary laws effectively excluded the poisonous marine animals of both the Mediterranean Sea and the Red Sea from the diets of the Israelites. The admonition in Leviticus regarding seafood—"all that have fins and scales ye shall eat; and whatsoever hath not fins and scales ye may not eat"—prohibited the consumption of, for example, puffers, fishes which are now known to carry large amounts of tetrodotoxin.

For these Red Sea glassy sweepers, the anonymity and confusion of large schools provide protection from predators.

In a similar vein, Chinese medical texts from the T'ang dynasty (618–907 A.D.) warned against the periodically harmful nature of *Seriola,* the yellowtail amberjack, described as a "large poisonous fish, fatally toxic to man." *Seriola* is one of the many Pacific marine fishes now known to be potential carriers of ciguatoxin.

Both tetrodotoxin and ciguatoxin are powerful nerve poisons that interfere with certain pathways in cell membranes through which ions move in and out of cells. Nerve cells use movement of charged ions back and forth across their membranes to carry messages through the body and as a result are especially sensitive to these poisons. Problems with the conduction of nerve impulses and the transmission of impulses between nerves and from nerves to muscles give rise to varying combinations of symptoms, including cramps, muscular pain and paralysis, loss of sensation, and sensory hallucinations.

Tetrodotoxin, first isolated from puffers of the family *Tetraodontidae,* is one of the best-known marine poisons. It consists of a single chemical whose composition is well known, although debate about its structure continues. Most body tissues of many tetraodontid fishes contain some tetrodotoxin, and several internal organs may be saturated with high concentrations of it at certain seasons. Adult puffers rarely need tetrodotoxin's protection; they can make themselves unattractive to most predators by inflating their bodies and erecting their spines. Larvae and very young puffers, however, lack protective spines. The particularly high concentration of tetrodotoxin in puffer ovaries and eggs may help protect these vulnerable juveniles. Predators apt to devour young puffers can probably learn from a few uncomfortable experiences that these particular species are not to be tampered with.

People still regularly die from tetrodotoxin poisoning, despite sound knowledge both of its danger and of the species that carry it. As early as the sixteenth century, a Japanese researcher tested the toxicity of various puffers by feeding puffer flesh and internal organs to human prisoners. (Nowadays, scientists use pigs, mice, or mongooses in toxicity studies.) Nonetheless, several species of puffers—known as fugu by the Japanese—are highly prized by Oriental gourmets, who pay more than $25 a pound for these fishes. When properly cleaned and eaten raw as sashimi, the meat from most fugu species is considered "relatively safe," but other widely appreciated preparations involving the internal organs entail greater risk. Fugu poisoning is nothing to be trifled with; Halstead quotes Englebrecht Kaempfer's early eighteenth-century *History of Japan,* which relates how

> five persons of Nagasaki having eat a dish of this Fish, fainted soon after dinner, grew convulsive and delirious, and fell into such a violent spitting of Blood, as made a violent end of their lives within a few days. And yet the Japanese won't deprive themselves of a dish so delicate in their opinion, for all they have so many Instances of how fatal and dangerous a consequence it is to eat it.

Over 250 years later, and despite careful training and government licensing of fugu chefs in Japan, several people still die each year from ingesting too much tetrodotoxin. Despite the danger, fugu fanatics still relish the mild euphoria, hot and cold flushes, and tingling sensations afforded by "controlled" doses of the poison, as well as the unusual taste of the meat. As even sublethal symptoms of severe tetrodotoxin poisoning include sweating, headache, nausea, respiratory paralysis, skin rashes, hemorrhaging, deep apparent coma without loss of consciousness, and sometimes total muscular paralysis, it is surprising that sashimi fanciers do not stick to tuna and sea bass.

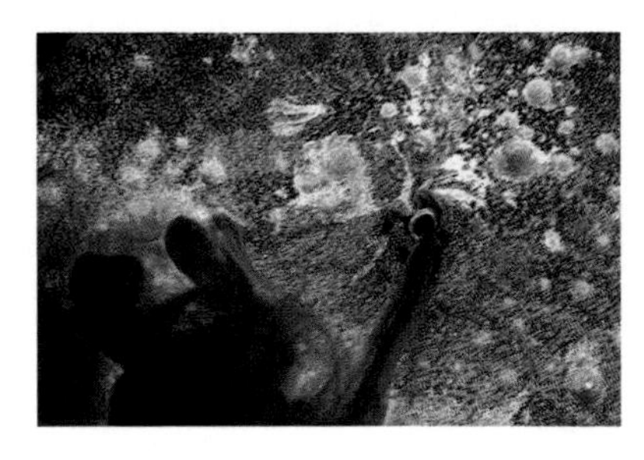

Many nudibranchs and other shell-less molluscs release toxic or noxious chemicals when disturbed.

Pufferfish and related species often contain the deadly poisonous tetrodotoxin in their flesh and internal organs.

Should any larger predator attack this scorpionfish lying in ambush, it will be most unpleasantly surprised by the powerful venom carried in the camouflaged dorsal spines.

The menace to people from the phenomenon of ciguatera, on the other hand, is due to its unpleasant propensity for cropping up unpredictably in locations—almost always in the tropics—and in fish species that have previously been perfectly safe. To make matters worse, the toxin is colorless, odorless, heat resistant, and insoluble in water, so soaking and cooking suspect fish do not remove it. Readers in the temperate zone might think it most sensible simply to avoid species known to be toxic, but Halstead's list of potentially ciguatoxic species runs to 65 pages and includes over 300 species! Symptoms of ciguatera poisoning include intense abdominal pains, nausea, vomiting, diarrhea, overall numbness, headache, anxiety, fear, fever, insomnia, severe exhaustion, and weakness, with aching or shooting muscle pains.

Today, well over 1,000 years since the first recorded occurrence of ciguatera poisoning, the nature of ciguatoxin is still not clear. The diversity of reported symptoms has led researchers to the conclusion that the poison is a complex of several compounds, varying in their proportions from place to place. Perhaps as a result of this variability, no dependable chemical analysis of ciguatoxin is as yet available.

Tracking down ciguatoxin's source has produced one of the most fascinating detective stories in the field of marine biology. Of course, no single ocean-going Peter Wimsey is responsible for solving the mystery alone; scientific endeavors of this type involve dozens or even hundreds of field and laboratory researchers around the world. But a notable few—Dr. John Randall and Dr. Albert Banner, both of the University of Hawaii, and Dr. Takeshi Yasumoto of Tohoku University, among the more recent ones—have assembled bits of scattered information and have tied them together into a coherent account.

For many years, ciguatoxic fishes were thought to be ill themselves, and some early writers labeled them rabid or "possessed of a distemper." Guesses closer to the mark soon began to appear, and during the colonization of the New World, one explorer—Peter Martyr of Anghera—correctly concluded that ciguatoxic fish eat something that afflicts people who later eat them with "divers strange maladies." Although Martyr was incorrect in guessing that ciguatoxic species pick up their toxicity from a species of seaside terrestrial berry, he was on the right track. It took several hundred years for scientists to identify positively the source in fishes' diets that carried the poison.

Only in the past few decades have sufficient clues accumulated both to incriminate the guilty organism and to explain how the poison accumulated in carrier fishes. Early information about ciguatera puzzled researchers as much as it helped. Investigators determined that most ciguatoxic species either live entirely on coral reefs or migrate there to feed, usually on the bottom in shallow water. Ciguatoxic species may feed on bottom-dwelling algae, detritus, or other fishes and invertebrates; in carnivores, the degree of toxicity is proportional to the amount of fish in their diet. A species may be toxic in one location but not in another on the other side of an island or even a few kilometers away. Safe areas may suddenly support a crop of toxic fish, and—over long periods of time—poisoned areas may become safe again. The most severely toxic specimens tend to be large; in one locality only 25 percent of the toxic species weighed less than 1 kilogram, but 75 percent of the species weighing over 5 kilograms were toxic.

Things began to fall into place when all of this information was combined with the observation that ciguatoxic fish populations often first developed after severe seasonal storms (such as monsoons), after irregular storms (such as hurricanes), after earthquakes, or after human-caused disturbances (such as dredging or the sinking of a large ship). It became clear that fish acquire ciguatoxin because of some introduced factor in their immediate environment, and that toxicity is somehow related to food supply. The organism in which the poison originates seemed to be bottom-dwelling and restricted to shallow water.

But here the problem became complicated once again. As both herbivores and detritus feeders may be poisonous, the source of the toxin could be an alga, a fungus, or a bacterium, many of which are jumbled together in a hopeless tangle on any underwater surface. Because the toxic organism is found only in shallow (and hence well-lit) areas and is eaten by strictly herbivorous fishes, an alga seemed a likely candidate. Yet the most poisonous fishes of all are large carnivores.

If the toxic organism were a plant, carnivorous species must have the ability to acquire the toxin and to concentrate it in their tissues. The mechanism for this concentration was experimentally proved by feeding large, originally nontoxic carnivores on flesh from mildly toxic herbivores. Over the course of the experiment, toxicity was clearly passed from prey to predators, showing that ciguatoxin travels from one trophic level to the next. Ingested ciguatoxin is neither digested nor eliminated, so it becomes more and more concentrated at each step along the food chain, from plants to herbivores to first-level carnivores to second-level carnivores. This makes large, top-level flesh-eating species—which are, almost without exception, the fishes preferred by people—especially deadly poison reservoirs. The toxin does not last forever, but it does persist in the flesh of fish that ingest it for long periods of time. If the source of the poison is removed, however, the fish's system will eventually clean itself up.

Having decided that the toxin probably originated in a benthic alga, researchers redoubled their efforts to identify the species and to isolate the factors leading to its proliferation after phenomena as diverse as rainstorms and shipwrecks. The latter phenomenon, which strongly suggested the leaching of a toxic "something" from sunken vessels into the food chain, gave rise to several investigations that led only to dead ends.

Ultimately it became clear that shipwrecks and heavy storms both make available clean, uncolonized, hard surfaces within reach of sunlight. Newly sunken ships clearly offer vast amounts of unclaimed hard surface on exposed decks, turrets, and railings. Fresh-water runoff clears space on hard bottoms by killing many bottom dwellers, including corals, either directly by lowering salinity or indirectly by covering them with silt.

This opening of space is the key to the problem. In all reef areas, virtually every inch of hard surface is covered with algae, corals, sponges, sea anemones, and so on—all engaged in a constant and vigorous struggle for living space that just happens to occur too slowly for human observers to recognize during a casual visit. Each of these organisms has its own techniques for preventing other organisms from settling on top of it or invading it from the side, but some have more effective defenses than others.

Whenever a patch of hard surface becomes available, a steady stream of spores and larvae colonize it, in the same way that seeds and juveniles of terrestrial plants and animals immediately begin to reclaim an abandoned field. Some rapidly reproducing species are adept at asserting dominance in the new territory almost immediately; others—a bit slower to colonize but more successful in long-term competition for space—move in later. The resulting process of steady replacement is called ecological succession. Researchers hypothesized that the organism responsible for synthesizing ciguatoxin was one of the early colonizers of clean, hard surfaces. A rapid bloom of this species would follow any perturbation that generated

These surgeonfish are typical of normally harmless herbivorous fishes that may carry ciguatoxin from time to time.

the proper environment for it. After a time, however, its numbers would decline as it was replaced by the other algae, sponges, and corals that normally inhabit less disturbed areas.

Aided by a clear focus, search efforts for the ciguatoxin-producing species were redoubled. After several false leads, the research team led by Yasumoto picked up a trail that finally led to a small dinoflagellate, a relative of the organism that causes the poisonous red tide of temperate waters. This dinoflagellate—an organism new to science—does not always grow directly on the bottom itself, but often clings to the surface of multicellular red or brown algae that are themselves early successional species in coral reef areas. The offending alga has been cultured in the laboratory and turns out to manufacture two poisons. One of these has been called ciguatoxin, and the other maitotoxin (after the native name for the fish from whose flesh it was first isolated). Maitotoxin is far more harmful to fishes but is unstable under heat and soluble in water, so boiling effectively eliminates its danger to people.

But scientists are still unable to predict accurately exactly when and where ciguatera outbreaks will occur, and they do not know how to tell whether fishes in an unfamiliar area will be toxic or not. Finally, it remains unclear what evolutionary advantage, if any, the ciguatera phenomenon offers any of the participating animals. Ciguatoxin does not seem to protect its algal manufacturers from herbivores (although maitotoxin might), and it does not appear to benefit herbivorous fishes that carry it either. It certainly fails to protect them from predators, and even in the extremely high concentrations it reaches in top-level carnivores, ciguatoxin seems not to affect the fish at all.

Ciguatera toxicity appears to be one of those accidents of nature with no evolutionary raison d'etre. The toxin, produced as an algal metabolic by-product, has a chemical structure that is nontoxic, indigestible, and nonexcretable by fishes. But perhaps there is a hidden lesson here after all. It is through fortuitous accidents that natural selection operates. Ciguatoxin—which does not appear to be toxic to fishes—arose at the primary-producer level and coincidentally becomes concentrated higher up in food chains through a process called biological magnification. Specific biological phenomena need not all be of adaptive significance; as long as a characteristic of an organism doesn't decrease that organism's fitness, it may appear and persist indefinitely. Besides, the question of evolutionary utility depends a great deal on perspective: if extraterrestrial biologists were to examine the current situation, might they not suspect that the entire system evolved to protect jacks and barracuda from overfishing by humans?

In contrast to the evolutionarily enigmatic ciguatoxin, other biologically active substances found in fishes have clear evolutionary significance. The shy, retiring flatfish known as the Moses sole (*Pardachirus marmoratus*) is second to none in its visual camouflage, which makes it extremely difficult for human observers to locate. Sharks that hunt *P. marmoratus,* however, supplement their visual search with an efficient sense of smell and the ability to home in on prey by following electrical signals generated during breathing movements. Thus, *P. marmoratus,* nearly invisible to humans and to vision-dependent reef predators, are eminently detectable by predatory sharks.

As a natural defense, the Moses sole manufactures a milky white substance called paradoxin and continuously releases it into the surrounding water from glands located at the base of each of its fin spines. In laboratory experiments conducted by Dr. Eugenie Clark, paradoxin froze attacking sharks in mid-lunge with jaws agape, torn between hunger and their aversion for the chemical. A combined Israeli-American research team recently

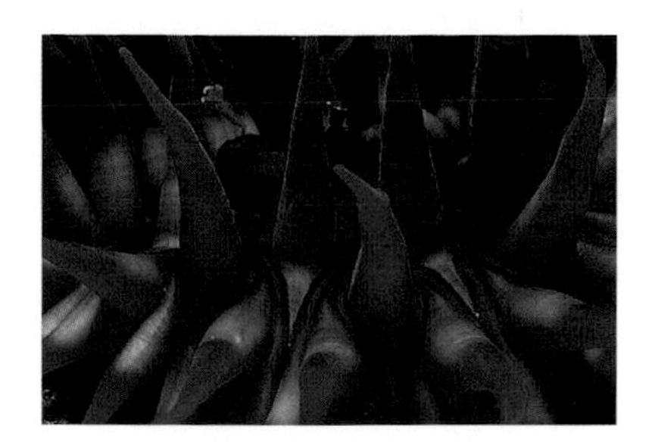

Tentacles of large anemones like this Californian species often contain stinging cells with venom strong enough to stun small fishes.

Following pages:
The stinging cells in these colonial anemones are dangerous only to the planktonic animals on which they feed.

determined that paradoxin alters delicate cell membranes in fish gills sufficiently to kill many other fishes and to discomfit sharks. Researchers actually found three active ingredients in the gland secretion, one of which acts as an inhibitor of the poisonous ingredient. This part of the secretion, which may explain the sole's immunity to its own poison, also inhibits the toxic action of several insect and snake venoms, making the Moses sole potentially useful as a medical tool and as a source of much-needed shark repellent.

It is tempting to characterize venomous animals as malignant demons eagerly awaiting hapless victims on which to pounce with toxic tooth and envenomed claw. But structures as complicated as venom glands and the biological mechanisms associated with injecting toxins—although they may resemble the diabolical weapons of man-eating monsters found in myth and science fiction—do not represent the whims of an evil intelligence. Rather, these structures exist because, as inconspicuous as their evolutionary origins might have been, they have somewhere along the line aided their bearers in surviving and reproducing.

Perhaps the most conspicuous marine animals that use venoms to capture food are corals, sea anemones, and other members of the phylum *Cnidaria.* All of these soft-bodied creatures possess batteries of stinging cells containing minute, barbed darts that lie tightly coiled inside egg-shaped venom sacs called nematocysts. Upon receiving the proper stimulus, nematocysts uncoil explosively and inject their venom into the prey they penetrate. For many years, researchers were unsure how nematocysts were fired; although cnidarians possess simple nervous systems, no evidence indicates that nerves ever attach to the stinging apparatus. All nematocysts exhibit trigger devices that seem designed to fire when touched, but contact alone does not always cause the nematocysts to discharge. Several animals that live symbiotically with cnidarians manage to snuggle up to the stinging tentacles with no apparent harm, and it would certainly do a cnidarian little good to attack the rocks and sponges surrounding it every time its tentacles brushed against them in the current.

By experimentally stimulating nematocysts with a variety of natural and artificial objects ranging from sterile glass rods to fish scales, researchers determined that nematocysts use a primitive sense of taste to distinguish potential meals from inanimate objects. All animals steadily release a variety of compounds into their environment, producing a kind of chemical fingerprint. By detecting the presence or absence of such fingerprints, nematocyst sensory structures differentiate between potential food items and inanimate objects. This simple sense of taste, located in the nematocysts' trigger hairs, does allow for certain exceptions: by using mucus to mask their skin compounds from the nematocysts, clownfish and several other fishes and invertebrates are able to live in close symbiosis with their otherwise deadly hosts.

Most nematocysts are designed only to capture small fishes or even smaller planktonic organisms, and are either too weak to penetrate far into human skin or are equipped with venom too feeble to affect creatures as large as ourselves. Many (but not all) jellyfishes and the vast majority of anemones and corals fall into this category. A swimmer touching one of these animals might find it somewhat "sticky" as a result of discharged nematocysts, but unless the barbs make contact with particularly sensitive areas such as the groin, breasts, or armpits, the swimmer will experience no pain.

An entirely different picture is presented by hydrozoans of the genus *Physalia,* the infamous Portuguese man-of-war. Hanging from the shim-

mering, pale blue, bell-shaped floats that are this genus's hallmark are nearly invisible stinging tentacles that trail out in the current for many meters. Since the loaded nematocysts require no instructions from a central nervous system to cause them to fire, free-floating tentacles broken off in storms or washed up on beaches remain dangerous for quite some time. *Physalia* nematocysts are armed with a potent neurotoxic venom that, in human beings, causes severe, sharp or shooting pain, followed by headaches, chills, shock, or even hysteria, depending upon the number of stings inflicted. Encounters with large numbers of tentacles can result in long-term incapacitation, and even death.

The severity of the threat from *Physalia* pales into insignificance, however, when compared with that presented by sea wasps of the genus *Chironex*. Sea wasp stings—normally used by the animals to stun and capture bite-sized prey—can cause intense, immediate pain, muscular spasms, and paralysis in swimmers unfortunate enough to brush against them. Pain from sea wasp stings may cause victims to drown before the venom itself does them in, but severe stings can be fatal by themselves. Even if the victim is rushed to a hospital, irreversible cardiac and respiratory paralysis often sets in within minutes. Fortunately for most people who spend time in the sea, the most dangerous species in this genus are restricted to a relatively small area of the tropical South Pacific.

Another familiar invertebrate group that includes some species dependent on venom to catch their prey is the phylum *Mollusca,* among whose subgroups are snails, clams, octopi, and squid. Although gargantuan octopi and squid (kraken) have always been favorite fodder for horror movies, real-life venomous molluscs are shy and retiring by nature and usually attack people only when provoked. Venomous octopi and cone shells use their toxins to paralyze and subdue a variety of small worms, fishes, crabs, shrimp, and other molluscs. When captured and handled indiscriminately, though, they exhibit little reticence in applying their venom to human beings.

Cone shells attack using a long, flexible proboscis tipped with an ejectable, venomous tooth that sinks easily into the flesh of their prey. The parrotlike beaks of octopi are accompanied by two pairs of venom-producing salivary glands that can either discharge venom into the water surrounding invertebrate prey or pump it into an open wound made by the mollusc's beak in a larger animal. The neurotoxic venoms of cone shells and of octopi have been known to kill people; severe neurological and neuromuscular distress accompanies many nonfatal bites.

Several other molluscs have unexpected defense mechanisms, as well. Several nudibranchs—marine snails that have evolved into shell-less animals—produce poison in several glands located within their fleshy bodies. This fact was well known in Aristotle's time, when the mere sight of a species of *Aplysia* was (erroneously) thought to cause spontaneous abortions in pregnant women, and touching one was believed fatal. Some nudibranch species release clouds of noxious secretions if disturbed, managing to hide within a visual and chemical smokescreen. Still other nudibranchs make use of ingested nematocysts for defense, in a manner analogous to the way several relatives make use of ingested pigments for camouflage. These animals feed on cnidarians and use special chemicals in their digestive tracts to prevent nematocysts from discharging as they swallow them. The stinging structures are then transported through the animals' digestive tracts to the tips of their gills, where they provide very effective protection from predators.

Cone shells carry venomous harpoons with which they can paralyze small prey and even sizable predators.

Armor and Armament

Physical armor supplements chemical weaponry among members of the phylum *Echinodermata,* common inhabitants of temperate and tropical seas around the world. Their jointed, five-part external skeletons are often covered with spines that can be either fixed solidly to the shell or joined flexibly to it by elaborate ball and socket joints that enable the animals to move spines around at will. Between the spines are flexible, inflatable tube feet that function in the same manner as suction cups, allowing both sea urchins and starfish to move around, and providing starfish with their renowned ability to pull apart clams and mussels.

Even stranger than the tube feet are the pedicillariae – sets of two, three, or four fanglike jaws located at the ends of flexible stalks. These peculiar structures help starfish break up and remove settling organisms that might try to overgrow the animals' hard external skeleton. Hundreds of them, acting as tiny pincers, can also be pressed into service in the capture of tiny fishes unfortunate enough to rest against the starfish's hide.

Sea urchins are ardently sought after as entrées by marine species ranging from triggerfish to several of the sea urchins' own relatives, the starfish. Spines – some venomous, others not – are sea urchins' primary defense against predators. People should approach all sea urchins with care, too, since even spines that lack venom are brittle and covered with thousands of backward-pointing barbs. Easily broken off once they enter the flesh, these barbed punji sticks can work their way deep into muscles and joints, causing long-term irritation. Even sea urchins' spines, however, do little to deter predators with jaws as powerful as those of parrotfish and triggerfish. To avoid these well-adapted but strictly diurnal predators, sea urchins use their spines to wedge themselves into coral crevices by day, emerging to forage for algae along the reef as twilight approaches.

Some sea urchin species have venom glands in both their spines and their pedicillariae that continue to inject venom even after they have become separated from the animal. The pedicillariae in these species rush to attack the tube feet of predatory starfish, a presence they detect chemically.

Stingrays – perhaps because of their close evolutionary relationship to the much-maligned sharks – have acquired particularly unsavory reputations dating back to the time of Pliny the Elder (first century A.D.). Pliny is credited with the assertion that

> there is nothing more to be dreaded than the sting which protrudes from the tail of the Trygon, by our own people known as the Pastinaca, a weapon 5 inches in length. Fixing this in the root of a tree, the fish is able to kill it; it can pierce armor too, just as though with an arrow, and to the strength of iron it adds all the corrosive qualities of poison.

Setting aside the question of why a ray should flop out onto a wooded shore and drive its spine into tree roots, one can see that this weapon has long been considered a thing not to be trifled with. After long debate on the uses to which rays normally put their spines, workers in the late 1940s found an answer. Several species of sharks known to feed on skates and rays were examined, and researchers discovered stingray stings embedded in the jaws, throats, or elsewhere around the heads of these sharks. Small wonder popular myths describe rays as vicious!

Most of the fear of rays generated by these stories is unwarranted, but these fishes are capable of inflicting serious wounds on people. The barbed stings normally kept sheathed on these animals' tails can penetrate deeply into human limbs, ripping gaping holes lined with venom as they are pulled out. The truth about stingrays, though, is

The thick shells of these moon snails provide safe retreats into which the animals can withdraw if threatened; they spread their fleshy feet outside to feed and to mate.

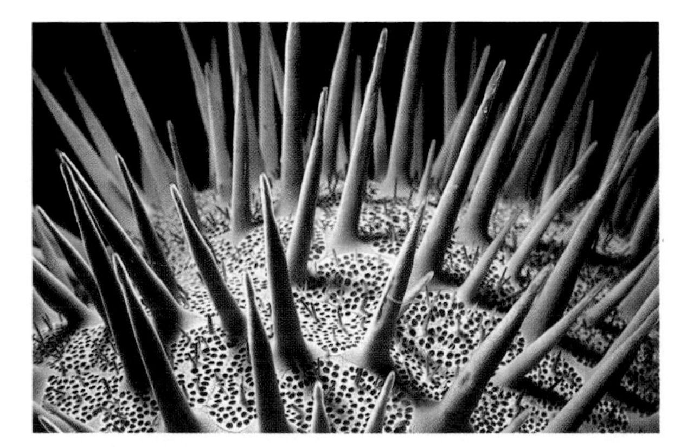

Crown-of-thorns starfish spines carry a mixture of mucus and bacteria that can cause painful swelling in the puncture wounds they inflict.

that they are not particularly aggressive nor particularly active. Rays lash out at people with their tails only when they feel imminently threatened—such as when stepped on.

As members of one of the oldest groups of living fishes, rays have been around since at least the time of the dinosaurs and have improvised in unexpected directions. One genus, *Torpedo,* has evolved a means of defense that also serves as a prey-catching mechanism. All nerves and muscles in animals' bodies generate small electrical impulses, but since these impulses are randomly oriented, they rarely amount to anything people can detect without the aid of electronic amplifying devices. In *Torpedo,* however, all of the cellular components of whole muscles line up in such a way that the electrical potentials they generate unite just as do those of batteries placed in series. When scores of these cells fire an impulse in unison, they create an electric shock more than sufficient to stun a person, and certainly adequate to discourage enemies and paralyze small prey.

Quite a number of the more advanced, bony fishes sport venomous spines that they keep concealed until needed as defensive last resorts. Tangs and surgeonfishes, members of the family *Acanthuridae,* sheath razor-sharp spines in natural scabbards on either side of their tail base. These generally timid creatures are quick to flee approaching danger, but they change their manner rapidly if cornered, hooked, or speared, baring their spines and fiercely slashing at attackers.

Two groups of commercially valuable food fishes, the weaverfishes and the rabbitfishes, carry various combinations of dorsal, pectoral, and gill-cover spines impregnated with venom that make the wounds they inflict far more distressing than the fishes' small size would indicate. Neither weaverfishes nor rabbitfishes normally attack humans, but when caught in fishing nets or on the ends of hooks, they thrash and twist violently, exposing their spines for full effect. Rabbitfish wounds are painful, but the agony that follows weaverfish stings is legendary among fishermen from the Mediterranean all the way to Scandinavia. There exists at least one documented case of a man who actually amputated his own finger in a desperate attempt to rid himself of the excruciating pain of a weaverfish sting.

Probably the most spectacular and bizarre of venomous fishes belong to the family *Scorpaenidae,* whose most dangerous members are restricted to the Pacific. Venomous scorpaenids can be divided by appearance into two types: the frilled and lacy lionfish and turkeyfish, and the magnificently camouflaged stonefish and scorpionfish.

Lionfish are among the most ornate fishes to be found on the coral reef. All of their long, curving, venomous fin spines are covered with delicate tissue that ripples as the fish glide through the water. Lionfish do not normally use their venom when capturing prey; they simply use their fan-like pectoral fins to corner their victims before making a sudden lunge and a split-second gulp that sucks in their prey faster than the eye can follow. They can, however, respond quickly to humans or other perceived enemies if they feel threatened, erecting their dorsal spines and making quick, jabbing movements in the direction of the threatening object. Lionfish venom combines neurotoxic action and powerful hemolytic activity that destroys blood cells in massive numbers and causes serious tissue damage.

While all divers maintain a healthy respect for lionfish, it is stonefish and scorpionfish that they fear the most. Unlike the flamboyant lionfish, stonefish sit quietly ensconced amidst corals and algae or partially buried in soft sand. These ambush hunters possess extraordinary ability to mimic the features and colors of their surroundings, which enables them to hide from potential prey until the victims are within gulping distance.

The near perfect camouflage of these species completely conceals the entire fish—including a razor-sharp set of dorsal spines that bear deadly neurotoxic and hemotoxic venom. If they sense danger, stonefish erect their dorsal spines and attack, jabbing repeatedly at intruders.

For scientists working with poisonous and venomous organisms, though, the current lack of knowledge about biological toxins is at least as distressing as the physical threat from toxication itself. There is, to be sure, a great body of knowledge on the subject: *Poisonous and Venomous Marine Animals* is about 1,300 pages long. The information it contains, though, is spread out thinly over hundreds of toxic species from all the world's oceans. Halstead's book is peppered with phrases like "No knowledge is available on the nature of this toxin," "No antidote is known for this poison," and even "Pathology, Chemistry, Toxicology—unknown." Modern Western science has been able to manufacture antivenoms for only a few of the more potent venoms of the Indo-Pacific area, and effective antidotes for poisons are few and far between.

Native Pacific islanders, however, have conducted their own research for centuries and have accumulated a sizable list of remedies for stings, poisons, and venoms, based on a variety of medicinal plants. Only recently have more than a handful of Western scientists realized that, just as these people often know best which species are poisonous and when, they may also be on to something in the way of remedies. Researchers from Japan, Australia, and the United States are now conducting chemical analyses of many plants used in island bush medicine to determine which might yield useful curative chemicals. Similarly, although Western science has not yet discovered a simple, reliable test to indicate the presence of ciguatoxin in newly caught fish, some islanders claim to be able to detect when the poison is present by observing the behavior of flies around suspect meat. These and other "local superstitions" are finally getting the serious attention they deserve.

Some Western researchers, too, have realized that medicine has many potential undersea allies in its continuing war against human parasites and disease. Every year, more and more disease organisms evolve strains resistant to the limited, land-based supply of antibiotics and other medications, which seriously concerns both physicians and drug companies. Although pharmaceutical research has held the line against many diseases thus far by exhaustively searching terrestrial organisms for new antibiotic compounds, their stock of new drugs is running low.

Poisons and other compounds, such as atropine, belladonna, and digitalis, that are obtained from terrestrial plants have been invaluable medical aids. Yet biomedical exploration has hardly scratched the surface of what promises to be a treasure trove of useful biocompounds from the sea. Marine organisms from horseshoe crabs to sharks to sponges to soft corals have perfected chemical defenses based on viricidal, bactericidal, and fungicidal compounds of which we know but a few. Tetrodotoxin has already come into common use in studies of nerve-cell membrane function. The discovery of paradoxin has opened new avenues in antivenom research, as well as providing an opportunity to develop a commercially viable shark repellent. Compounds extracted from sponges and soft corals may prove useful in treating infections ranging from sore throats to herpes and even certain forms of cancer. Directed research in marine biomedicine, in combination with conservation efforts to preserve marine organisms and their habitats, thus has the potential to tie living systems of land and sea together in yet another manner, adding a few more interconnecting strands to the global network of life.

Lionfish carry potent venom in spines that trail from dorsal and pectoral fins.

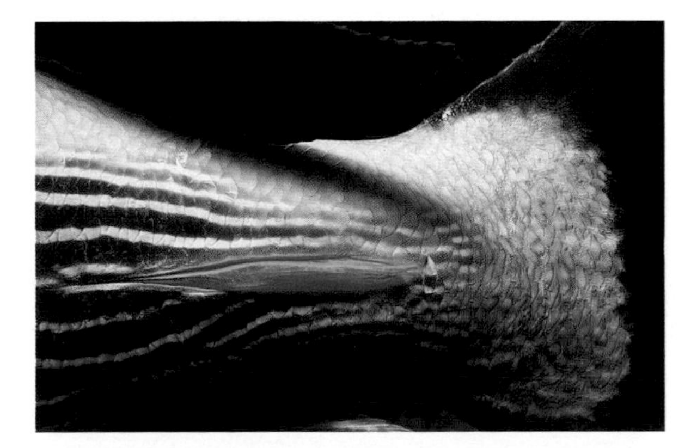

Used only as defensive last resorts, surgeonfishes' razor-sharp "switchblades" can slash deeply into the flesh of humans or other potential predators.

To Exhaust the Sea

Ocean's Bounty

A cold, northeast wind whips across the deck of a New England trawler that rides the North Atlantic swells while a straining electric winch hauls in a netful of dripping, thrashing fishes. Two hours of dragging a weighted net along the sandy bottom has yielded a full load of flounder, pollock, sea raven, sculpins, goosefish, sand sharks, skates, and even a few lobsters and octopi. Sorting out food fish from "junk fish," the crew packs the catch in ice within plastic-lined boxes and stows these in the hold until the boat returns to port in the evening.

Half a world away, a dry desert breeze ruffles the sail of an Egyptian fishing boat, coaxing the vessel slowly through the calm, tropical waters of the Red Sea. Its mast creaks periodically, its seldom-used diesel motor remains quiet, and waves lap gently along its hull. The crew fishes patiently, trailing hand lines baited with squid over the side in search of grouper, parrotfish, jack, and snapper along the coral reef.

In laboratories scattered through the northeastern United States, white-coated technicians process horseshoe crab blood to extract a compound called Limulus amoebocyte lysate, or LAL. LAL, valued at $15,000 an ounce, will be mixed with samples of drugs and human blood products in a critical test for dangerous bacterial poisons called endotoxins. Endotoxins, which surround human beings constantly, are harmless on skin and even in intestines, but if they enter the bloodstream, even in minute quantities, they can cause severe illness or death. If as little as a billionth of a gram of endotoxin is present in a milliliter of fluid, however, it will react with LAL to form an easily observable precipitate. Screening with this extract from horseshoe crab blood enables technicians to detect endotoxin contamination at far lower concentrations than they could easily detect by other means.

Human interaction with the ocean and its resources can have many different faces. Some fishermen in Third World countries explore new fishing technology provided by their governments, while others rely on native practices that predate written history. Commercial operations in high-tech societies in the Western world continually develop new aquaculture techniques suitable for competing within their economic systems, and researchers comb the blood and body fluids of marine organisms for medically and industrially useful compounds. In the meantime, nations such as Japan and China expand and develop traditional aquaculture methods, acquiring critically needed proteins, vitamins, and trace elements from an eclectic assemblage of marine organisms they have long included in their diets.

This diversity of interaction is the key to

Flounder squirm in the net that collected them from the sea floor off New England.

understanding both the promise of marine food production and the problems that accompany it. Subsistence fishing, quaint and old-fashioned by Western standards, has supported populations around the world for millennia and has consistently maintained the health of the fisheries on which it depends. Western-style fishing industries, on the other hand, have, in separate encounters, taken less than a hundred years each to exterminate such animals as the Stellar's sea cow, to drive whales to near-extinction, and to place many commercially important fish stocks in real jeopardy.

Fisheries' problems around the globe reflect differences not only in scale but in culture and in approach to natural resources. The island peoples of Indonesia and Polynesia, for example, have lived for centuries on isolated bits of land in the vast Pacific. Faced with the limitations of their finite terrestrial habitats, island cultures developed strong conservation ethics. Nature worship became intertwined with social taboos, and rules of conservation safeguarded the little land they had. In some cultures, disregard for nature-protecting customs was punished as severely as social crime.

The care these people took in protecting terrestrial resources carried over to their interaction with the sea. They learned that—despite the immensity of the ocean around them—many fishes they preferred as food lived only on and around island coral reefs. Islanders thus became part of their reef ecosystems, the way indigenous peoples on continents integrated themselves into the ecosystems of forests and savannas.

Islanders learned the patterns of tides and currents associated with lunar cycles. They observed fish feeding, schooling, and migrating. They located spawning sites and pinpointed reproductive seasons. With enviable foresight, they collectively established and observed closed seasons and catch limits to ensure the health and vigor of their living resources. Through these efforts, they maintained a steady yield that supported substantial, stable human populations until their islands were discovered by Europeans.

The fishing industry in Western countries, on the other hand, has tended to operate from implicit assumptions that the ocean is vast, its currents mysterious, the comings and goings of its denizens unpredictable, and its fish stocks heaven-sent and inexhaustible. This view dates back at least to the Middle Ages and persists despite ample evidence to the contrary. Throughout the medieval period, fortunes—not only of individuals, but of entire cities in the Hanseatic League—depended heavily on the relative abundance of herring stocks in the Baltic Sea and neighboring Atlantic areas. No one, however, seems to have considered that changes in catch might be related to human fishing activity rather than divine whim. Not until the middle of the nineteenth century did talk of marine fishery management even begin in England, for example, and the first limits on catch size were not recommended until the mid-1890s.

Today, many fishing fleets still operate with this medieval mentality, though by now fishery science is advanced enough that they should know better. Popular science writers still discuss the potential of "food from the sea" in glowing terms, implying that the oceans are somehow magically able to produce unlimited food.

Considering this world view, it is hardly surprising that Western fishermen concentrate, not on preserving the overall numbers of their prey, but on exploiting them to the maximum. Economic factors exacerbate the problem. Modern fishery fleets require large capital investments and necessarily focus attention on abundant species. When a fledgling fleet first sets out to sea, it reaps an ample harvest that nonetheless is so small relative to the vast swarms of its previously untapped target species that its fishing activities hardly make a dent in the population. For the first few years, the more time the fleet spends fishing and the more vessels join the fleet, the greater the catch becomes. Eager venture capitalists increase investments in fishing gear under these circumstances, and for a while, catch and profit from the venture continue to grow.

At such times in fishery development, large and small operators alike are united in their distrust of anyone interested in studying their target species and in their resistance to any group that recommends managerial restraint. And as long as the fishery as a whole remains small and the total catch insignificant in comparison to available stock, this system works. But if the fleet keeps growing, catch size approaches and eventually matches the population's ability to replace itself. If the catch size then exceeds the population's ability to replace the harvested individuals—a condition called overexploitation—the stock begins a sharp decline from which it may or may not recover.

It is not easy to predict overexploitation. Like spawning failures due to estuarine habitat destruction, visible results of overfishing often lag behind the events that cause them by several years. A fleet, therefore, may continue to catch adults for several years after overfishing has caused major reductions in the numbers of younger individuals. The first small catch is likely to be attributed to a "bad year." By the time the next reduction in mature individuals is noticed, the population may have become sufficiently small that it requires several years' respite to replenish itself—assuming that it has not already been reduced below the minimum density at which its members can effectively reproduce. When the latter occurs, extinction is the probable outcome.

But even if supervisors read signs of overfishing correctly, fishing fleets are not simply machines that can be shut down and left idle. Families of fishermen must be fed; mortgages on boats and equipment must be paid; investors expect dividends.

Instead of backing down in the face of reduced yields, fleets fish longer and longer hours, desperately attempting to maintain the catch in the face of declining stocks. At this point, major industrial fleets, which are often able to survive lean times by diversifying their catch, may grudgingly submit to regulatory authority. Public relations departments in big firms may even openly embrace conservation measures and feature their companies' compliance with them in media campaigns. But small operators who lack such flexibility may become bitter as they are forced toward bankruptcy.

As the catch of original target species finally falls below profitable levels, big fleets buy new equipment, switch targets, and begin a fresh cycle with a new species. Some large corporations, banking on the speed with which the public forgets the too-little-too-late character of their compliance with stock management techniques, even argue that their recent conservation-minded outlook should exclude them from externally imposed regulations in the future.

This pattern of overexploitation is not a new one; it has been repeated time and time again throughout the world, though the specific details of each story differ. Overexploitation of marine resources in the New World dates back to the very earliest European settlements, when rivers, estuaries, rocky shores, and coastal waters teemed with fish and shellfish. Salmon fought their way up eastern rivers. Cod, haddock, and pollock swarmed offshore. The Atlantic and the Pacific supported huge populations of whales. As for our rare and highly prized lobster, William Wood's early eighteenth century "true and lively" description of coastal New England included the following:

> Lobsters be plenty in most places, very large ones, some being twenty pound in weight. These are taken at low water amongst the rocks. They

Sea cucumbers and a host of other marine invertebrates are potential gold mines of antibiotics, viricides, and anti-cancer drugs.

are a very good fish, the small ones being the best; their plenty makes them little esteemed and seldom eaten. The Indians get many of them every day for to bait their hooks withal and to eat when they can get no bass.

Before the end of the nineteenth century, most salmon runs up eastern rivers had been eliminated. By the mid-twentieth century, lobster populations had been so decimated that their scarcity triggered ecological changes in northeastern kelp communities.

Whales are particularly susceptible to overexploitation because they take many years to mature and reproduce, and even after maturity they reproduce at very low rates. The Atlantic right whale, so named because it was the "right" whale to catch because it floated after harpooning instead of sinking, had been hunted almost to extinction by the end of the seventeenth century. Atlantic whalers then switched their attention to the Greenland right whale, with the same result.

By the early 1930s, it was clear that whales and the whaling industry would soon be extinct. The International Whaling Commission was formed to set protective quotas. Unfortunately, the IWC, like most international maritime organizations, had no real legal clout and often yielded to economic and political pressures to set quotas higher than fishery biologists recommended. Unable to enforce even these inadequate restrictions, the commission watched as member and nonmember nations alike violated its rules, and whale populations continued to decline precipitously.

The commission then recommended a complete moratorium on whaling to take effect in 1986. Probably too little and too late to save the blue whale from extinction, the moratorium cannot be legally enforced in any way and has already been denounced by such important whaling countries as Japan, the Soviet Union, Norway, and Chile, which have announced their intention to continue whaling. In light of the IWC's legal impotence in such matters, continuing pressure from public opinion and the activities of aggressive conservation groups such as Greenpeace may be the whales' only chance.

As tragic as whales' plight may be, it is far from uncommon. Many marine species, less intelligent and less apt to catch and hold public sympathy, face similar situations around the world. Pacific sardines once swarmed off the California coast by the millions, supporting a fishery that harvested over 500,000 tons a year in the early 1930s. But by the end of that decade, overfishing and environmental changes had caused a steady drop in population size. A carefully managed fishery might have weathered these problems, but overfishing continued despite observations that the stock was overexploited. By the mid-1950s, the industry was dead.

Apparently unable to learn from their northern neighbor's mistakes, South American fleets made a virtually identical series of blunders in exploiting anchovies off the coast of Peru. There, westerly trade winds create massive upwellings that combine with a cold, northward current to carry abundant nutrients into the photic zone offshore. Large phytoplankton grow densely enough to make plankton nets smell like new-mown hay. These large phytoplankton can be eaten directly by such large planktivores as anchovies, making a short, highly efficient food chain possible.

Peruvian anchovies originally supported not only human fishermen but countless seabirds and marine mammals. From local beginnings, the Peruvian anchovy fishery grew rapidly through the late 1960s, when factory ships harvested more than 11 million tons of anchovies annually. In 1971, this single fishery, in an area totaling a mere two hundredths of a percent of the world's ocean area, accounted for more than 15 percent of the total

Giant lobsters, once commonplace in New England, have become increasingly rare due to overfishing.

world fish catch. Peruvian anchovies became a major source of inexpensive fish meal, a high-protein food for livestock in wealthy nations around the world.

But the nutrient-laden currents are not invariable. Each year in late December, a warmer stream of water flows southward, interfering with normal circulation patterns and temporarily suspending the upwelling. Peruvian fishermen have known of this seasonal current since the nineteenth century, when they named it *el niño Jesús* (the child Jesus) because of its occurrence near Christmas. By the latter part of the century, biologists realized that this current varied in strength: every few years, it was exceptionally strong. Scientists adopted the name El Niño to describe these major episodes alone.

Soon after the southward current appears, the upwelling falters, nutrients for primary production decline rapidly, phytoplankton growth slows, and shock waves travel upward through the food web. Larval anchovies suffer particularly because at the time they exhaust their yolk sacs and seek their first meal they must find plankton of the right size and density or they will die.

In the early years of the anchovy fishery, the population was large and robust enough to recover from the periodic loss of its young with little trouble. A major El Niño in the late 1950s caused no noticeable effect on the fishery. By the time a strong El Niño occurred in the mid-1960s, however, the fishery had grown to nearly 10 million tons annually, and the catch fell markedly.

Fishery managers redoubled their efforts in order to offset lower yields and began taking juvenile fish in large numbers. When the next El Niño arrived in force in 1972, the catch dropped precipitously from 11 million tons to less than 2 million tons. Still they fished, perhaps falsely encouraged by the stock's slow recovery over the next few years; but another El Niño in 1976 sent the year's catch plummeting to less than a million tons. Soon, the great anchovy fishery was dead.

Eleven million tons of protein is not something the world can lose lightly. The loss of such vast quantities of inexpensive livestock feed placed unexpected pressures on Western agriculture, adding another sudden price increase to economies already suffering from an energy crisis.

Ecological effects quickly moved up the food chain in natural systems too. Whales, dolphins, and seabirds—all of which feed heavily or even exclusively on anchovies—virtually disappeared from the region. Many migrated up to 1,000 miles away to find food. Others, unable to travel that far, stopped reproducing. Many starved. The combined population of cormorants, boobies, and pelicans along the Peruvian coast was estimated at over 27 million in the early 1950s. Eating tons of anchovies yearly, these birds excreted large amounts of phosphorus-rich guano at their nesting sites. Over the years, these deposits accumulated up to 50 meters deep in places—enough to support an entire fertilizer industry. During the great anchovy crisis, however, bird populations dropped to around 4 million and never fully recovered. With their food supply severely depleted, only about a million guano birds survived the 1972 El Niño.

Since that fatal time, El Niño's periodic appearance has caused major disturbances in the local ecology, testimony to the sorry state of the remaining anchovy stock. The strong, several-month-long El Niño of 1982–1983—in addition to producing climatic changes all over North and South America—caused marine mammals and birds off Peru to disappear en masse once again. By November 1983, some had returned, although their numbers were far smaller than normal. Whether the missing animals had permanently changed their living quarters—a supposition not supported by unusually large sightings elsewhere—or had starved is not known as yet. How long

it will take for the normal food web in the area to reestablish itself is also a matter of conjecture.

Again and again this pattern has been repeated. Soviet and Japanese factory fishing fleets, having depleted stocks close to home, cruise the world ocean, repeating their mistakes elsewhere. The Atlantic menhaden fishery collapsed throughout the northern part of its range at about the same time the Peruvian fisheries crashed. A newer industry based on menhaden in the Gulf of Mexico is—at this writing—still in its growth phase; there, as elsewhere, the industry vehemently resists all efforts at setting quotas to prevent yet another crash. Worldwide stocks, both of herringlike fish for meal and oil production and of larger species used directly for human consumption, are being pushed to their limits in location after location, from the continental shelf of the United States to the stormy North Sea.

Some encouraging signs have emerged that governments and industries may be changing their attitudes toward living marine resources. Many countries have recently claimed the waters within 200 miles of their coastlines as Exclusive Economic Zones (EEZs). Within each EEZ, the nation in control asserts the right to regulate access, both by its own local fleets and by long-distance fleets of other nations. Having the exclusive right to exploit these coastal zones—which contain up to 95 percent of the world's fishery resources—encourages each nation to shoulder the responsibility for maintaining its resident fish stocks.

The EEZ system bypasses toothless international agencies and places control in the hands of individual nations that not only can set catch limits but are prepared to enforce them. Sovereignty over ocean resources produces conflicts, of course; the "tuna war" between the United States and Canada in the northwestern Atlantic is a typical case. Similar altercations among the United Kingdom, Denmark, Sweden, and Norway in the northeastern Atlantic, North Sea, and English Channel have erupted over quotas, overlapping claims, and enforcement techniques. These disagreements periodically bring a halt to fishing, and may cause temporary shortages of fish for market, but they also give stocks a chance to recover from overfishing. The real problems are faced by fishermen whose investments and livelihood are at stake.

Even smaller units within EEZs offer promise for maintenance of stable, resident stocks of species such as the heavily fished American lobster. Coastal New England states enforce laws that protect breeding female lobsters and stipulate the minimum size at which lobsters can be sold legally. In response to these regulations and to the intense fishing pressure exerted on near-shore lobster populations, lobstermen have developed a territorial social system related to the fishery. Fishermen in Maine, for example, form tightly knit local organizations that stake out and protect private fishing grounds and unofficially control fishing.

These fishermen are well aware that it takes nine years for a lobster to reach the current minimum legal size, and only a year or two less for it to reach reproductive age. Those who have been in the business for years have seen the precipitous decline of near-shore stocks in their lifetimes. As a result, most of them respect and defend the management practices they develop to protect their catch. By acting in this way, lobstermen have, in fact, developed social customs not unlike those of indigenous fishermen elsewhere.

While individual countries squabble over who has control over EEZs, they must also determine just what kind of control to exercise. Fishery scientists are now developing improved sampling techniques and sophisticated computer models that enable researchers to estimate stock levels and project trends in fishery economics more accurately than ever before. New computer models include

Limulus, the ancient horseshoe crab, provides invaluable services to medical practice and research.

Preceding pages:
California gray whales and their relatives were once commonplace; many species are now at the edge of extinction (© Howard Hall).

randomly introduced safety factors to predict the effects of unforeseeable occurrences such as El Niño. Complex new analyses can also take economic factors into account in calculating exploitation costs.

But all the world's regulations, quotas, and management techniques cannot help fisheries grow at rates of 5 to 7 percent annually as they did in the 1950s and 1960s. Data on most commercially important fish species—from anchovies to cod and pollock—indicate that current catches of these species are near their upper bounds. In fact, the catches of many commercially important species have already peaked, and several are declining. This sobering fact emphasizes that the ocean, despite its size, has only a finite ability to produce the foods people desire.

Some of this limitation stems from the dependence of most marine food chains on microscopic primary producers. On land, meat producers favored by human beings graze directly on such primary producers as grasses, providing meat for human consumption on the second trophic level. Fishes such as cod, tuna, and salmon, on the other hand, feed three, four, or even five trophic levels away from the phytoplankton that capture the sun's energy. Three additional steps in the chain decrease the percentage of primary production from this source available to people by a factor of over 1,000. Consequently, even the highly productive continental shelves and upwelling regions can provide sustenance for only a finite number of fish.

The real root of the problem people face in extracting food from the sea lies in the prehistoric frame of mind with which Western society views living marine resources. Human food producers on land switched from hunting wild game and gathering wild plants to herding and farming thousands of years ago. Herding and farming are so much more efficient than hunting and gathering that domesticated terrestrial plants and animals now provide twenty times more food worldwide than those gathered and hunted in the wild. But in the sea, with few exceptions, Western countries still catch over fifty times more free-roaming fish than they raise in aquaculture ventures, and they gather three times as much algae as they farm. Modern fleets use high technology to locate and process wild seafood, but they still utilize that technology to serve a hunter-gatherer mentality.

Governments and corporations are only now awakening to the fact that the ocean's store of wealth is not bottomless and that harvesting techniques and strict catch limitations are required in light of both ocean productivity and economic considerations. Decision-makers are wrestling with the need to transform human exploitation of ocean resources from a childlike, catch-as-catch-can approach to a more mature and studied plan that includes careful exploration of alternatives.

In evaluating alternatives such as aquaculture, the Western world has much to learn from several Asian and Indo-Pacific nations that have aquacultural experience sufficiently old to be firmly rooted in folklore. If oral histories can be accepted at face value, aquaculture in China began as many as 4,000 years ago, about the time silkworms were first cultivated. Fan Li, the legendary Chinese Croesus, is credited with writing the first treatise popularizing fish-culturing in 460 B.C., and the Chinese carried their aquacultural techniques with them as their merchants and influence spread across Southeast Asia. From early beginnings as a strictly local subsistence activity, aquaculture in Asia and the Indo-Pacific has grown into a vast commercial enterprise. In 1980, aquaculture provided the Chinese with more than 40 percent of their fishery products, and supplied Java with more than 60 percent of the fish consumed on that island. The world leader in commercial aquaculture is Japan, where Western-style industrial capitalism merged with traditional aquacultural techniques over a century and a half ago. All told, more than 75 percent of the world's aquacultural output comes from the Indo-Pacific region.

Aquaculture in one form or another has been practiced in Western nations for many years, too. The Romans cultured oysters 2,000 years ago, and Americans have been managing oyster stocks to some degree since the 1850s. But aquacultural output in Western countries has never approached the level in the East. This can be explained in part by idiosyncrasies of public taste. Despite the ease with which it can be cultured, the mussel is shunned by most Americans and so is not commercially viable. From the standpoint of taste, the lobster would seem ideal for aquaculture; yet certain characteristics—they are aggressive, cannibalistic, and fussy about water quality—combine to make raising them highly labor-intensive.

More important, perhaps, are the social and economic requirements of food production, which create entirely different sets of problems for would-be aquaculturists in the United States and Europe than are faced elsewhere. Subsistence aquaculture is designed to produce edible or agriculturally valuable organisms for local consumption or use. Fish farmers in the Third World, therefore, can earn a satisfactory living by growing organisms in low densities on a small scale. But entrepreneurs in developed countries start fish farms to turn a profit, a goal that requires high concentrations of organisms to be raised, high growth rates, and dependable means to control production. Western efforts at marine aquaculture, therefore, face many special obstacles that can only be overcome through long-term research efforts, with substantial investments of time and expertise.

The very nature of water as a growing medium makes concentrated culture of aquatic organisms substantially more difficult than husbandry of terrestrial organisms. Crowded cattle yards and chicken farms generate plenty of wastes, but uneaten food and accumulating manure do not foul the animals' air. Extra food and solid waste do remain suspended in water, however, encouraging bacterial and fungal growth that depletes dissolved oxygen and endangers the health of the organisms being cultured. At the same time, liquid waste can seriously damage the gills of fish and shellfish. Economically viable aquabusinesses on the grand scale preferred by major venture capitalists are therefore far trickier to manage successfully than either subsistence fish farms or terrestrial agribusinesses. These problems make aquaculture today a highly speculative investment.

Well-controlled husbandry operations also require dependable supplies of new organisms and hence require the ability to breed and raise stock independent of natural populations. Here again, aquaculture is at a disadvantage. Newborn cows, sheep, and chickens, though more fragile than adults, look the same, act similarly, and—within a short time—survive on the same foods and in the same environment as their parents. The great majority of marine animals and plants, on the other hand, have larval stages that look nothing at all like adults, eat completely different food, and often require totally different aquacultural conditions. The difficulty of satisfying these variable needs—only recently overcome on a commercial scale—has further hindered efforts to raise productivity in aquacultural ventures.

On top of all this, most terrestrial farm plants and animals have been reared in captivity for hundreds or even thousands of years. Most have been selectively bred for at least a century to improve strength, yield, disease-resistance, and reproductive ability. Yet—because of their complex life cycles—marine organisms have only been bred in quantity in the last decade. During the many years of marine aquaculture in the Orient, virtually all eggs, larvae, or juveniles needed for culture were

Schooling fishes—such as mackerel, herring, and anchovies—are easily harvested by commercial trawlers and have been overexploited in many parts of the world.

Sea turtles have been hunted to the brink of extinction the world over.

caught wild. Until the last few years, therefore, most subjects of marine aquaculture have been unimproved wild stock, whose physiology, genetics, and disease characteristics were completely unknown. Aquatic veterinarians are steadily expanding the meager scientific understanding of aquatic diseases and parasites at an exponential rate, making it increasingly possible to reduce what are still unpredictable or uncontrollable outbreaks of disease in culture systems.

The infancy of aquaculture in the Western world means that the field simultaneously holds great promise for the future and poses weighty practical and philosophical issues about the status of marine plants and animals. Genetic manipulation of wild marine organisms—both through traditional selective breeding programs and through genetic engineering—can offer dramatic improvements over the quality of present aquacultural stock. Yet mass breeding of marine organisms poses its own set of problems. Already, the essential genetic diversity of dwindling wild salmon populations on the west coast of the United States may be threatened by the release of multitudes of genetically uniform, hatchery-raised strains. Fishery managers may thus have to face sooner than they expect the possible negative impact of replacing wild fish stocks with domesticated ones. Wild terrestrial grazing animals around the world have already been replaced by artificially maintained herds; are fishery biologists ready for a similar phenomenon in the sea?

While aquaculturists occupy themselves with problems of animal and plant husbandry, biochemical investigations of marine organisms are simultaneously providing new perspectives on the potential contributions of marine organisms to medicine and industry. Both defensive and offensive chemical weapons extracted from marine invertebrates show promise as antileukemia drugs, antibiotics, and antiviral agents. Sea cucumbers and their relatives contain holothurin, a compound that has shown signs of anticancer activity and that may also be combined with digitalis to treat coronary disease. Some benign marine bacteria are known to produce compounds that attract and encourage settlement of oyster larvae, making them potentially valuable in the rearing of such animals in aquaculture. Seaweeds provide carrageen and alginates that are used in everything from ice cream to plastic products. Knowledge of such potentially valuable compounds in marine organisms is still in its earliest stages of expansion. The opportunities for growth and development in this realm of marine biotechnology are legion, though they involve many of the same potential practical problems aquacultural efforts do.

Equally encouraging is the growing cooperation in research, trade, and marketing between Western nations and the Orient. The blending of old and new techniques and the establishment of trade channels between producers and consumers of aquacultural products ensure that the farming of the sea is entering a new phase of growth. Improved techniques of aquaculture, coupled with protection of critical marine habitats, have the potential to harness the oceans' productivity more reliably than ever before.

The Tainted Cup

SCENE 1:

The warm, clear waters of the Caribbean

A young leatherback turtle paddles its 90-kilogram body lazily along the surface, munching on jellyfish and other gelatinous zooplankton. Leatherbacks usually live for many years and weigh as much as 450 kilograms. This one will not. It mistook several floating plastic sandwich bags for jellyfish and swallowed them. In a few days its intestines will be blocked by a twisted mass of indigestible plastic, and it will die.

SCENE 2:

The cold, turbid North Atlantic

Beneath the surface, cod and pollock spawn. Each female releases between 200,000 and 1 million buoyant eggs that float near the surface. These eggs usually hatch into larvae that join the zooplankton to feed and grow before metamorphosing into their adult forms and joining their parents near the bottom. But this batch has encountered an oil slick from a broken tanker. Many eggs have become fouled with a mixture of petroleum hydrocarbons. Twenty percent of the cod embryos and nearly half of the pollock embryos will die before hatching—from five to ten times the mortality rate of uncontaminated eggs observed in the laboratory.

SCENE 3:

The placid waters off the Yucatán Peninsula

An offshore oil well has blown out of control and spews crude oil into the Gulf of Campeche. Where winds and tides carry the oil shoreward, it drifts into estuaries where it coats exposed mangrove roots and adheres to sediments in tidal creeks. For the next three days, oysters, other root-dwelling filter feeders, and many animals that burrow in the upper layers of sediment will die en masse. They will decay rapidly, and their soft tissues will disappear without a trace. A few weeks later, fouling mangrove roots will begin to blacken, and the trees most severely affected will begin to die. Beneath the surface, the oil will consolidate in the sediments, forming a dark, poisonous ooze that will remain, hidden from casual observers, for years after the spill.

One dead turtle or a few thousand damaged fish eggs do not endangered species make. Rarely, if ever, does a single oil spill anywhere in the world damage a healthy ecosystem beyond its ability to recover.

But several dozen dead turtles can make a big difference to a species whose numbers have been reduced from thousands to hundreds by excessive hunting. Repeated damage to eggs and planktonic larvae can adversely affect the reproductive ability of fishes already pushed to their limits by heavy exploitation. Repeated oil spills can have significant, long-term effects on the relatively few salt marshes and mangrove swamps left undisturbed by coastal development in both the temperate zones and the tropics. And whenever estuaries are adversely affected, populations of migratory animals that depend on them for breeding or nursery grounds begin to suffer.

Though knowledge of the sea and its ecosystems is growing rapidly, the ability of human beings to alter those ecosystems permanently is growing even faster. As human population grows, ecologically critical coastal areas disappear under bulldozers and coastal pollution increases exponentially. Within a split second of geologic time, the human species has changed the face of the continents and has begun to affect marine ecosystems.

The species of plants and animals in those ecosystems have evolved their shapes, colors, defenses, behaviors, and physiological characteristics through thousands of years of coexistence with one another and with their environment. Each is extremely well adapted to life in the world to which it is accustomed. But if that accustomed world changes too suddenly or too severely, the adaptations may become useless or even counter-productive to its continued survival. Ecological webs are only as strong as their weakest strands, and sudden changes in only a few plant or animal populations may have major repercussions elsewhere in the biosphere.

Until the current century, man-induced changes have had little overall effect on the open sea. Ecosystems are, after all, reasonably robust. A slight perturbation in the natural order may have existed for a while, but some plants absorbed the extra nitrogen and phosphorus, while bacteria and other microorganisms were able to break many of the poisons down into harmless components.

But today the perturbations keep coming harder and faster. While under temperature stress from the unnaturally warm water discharged from a power plant, an estuary is inundated with oil. Chemicals in coastal waters discourage the growth of more and more species. Organisms heavily exploited by people go first, but others follow. Significantly, the shallow continental shelves, which produce over 95 percent of our seafood at present and hold the only real promise for supporting marine aquaculture in the future, bear the brunt of civilization's impact on the sea.

People have always clustered near the shore, both for business and for pleasure, and they show no signs of changing this preference. By the year 2000, more than fifty cities around the world will contain more than 5 million people each. Half of those cities are located on estuaries, including seven of the ten largest—New York, Tokyo, London, Shanghai, Buenos Aires, Osaka, and Los Angeles. Outside the megalopolises, coastal resort development chews up salt marshes and mangrove swamps in the United States, while the "reclamation" of land for farming and the cutting of trees for lumber and firewood imperil mangrove swamps in developing nations throughout the tropics.

As many as 27 percent of American estuarine areas classified by the United States Fish and Wildlife Service as commercial shellfishing sites are closed because of pollution. The Caribbean coast of Mexico supports important fisheries and offers vast potential for producing medically and industrially useful compounds, but it is also the backbone of one of the busiest petroleum-related industrial areas in the world and is exposed to an ever-increasing number of oil spills and exploration-related disturbances. As industry runs out of terrestrial sites to use for disposing of chemical wastes, both corporations and governments look to coastal waters for convenient disposal sites. In Japan, for example, although shrimp and *nori* producers can never supply market needs, the pollution of shallow areas suitable for aquaculture hampers aquacultural expansion and in fact threatens the industry with slow decline.

Present knowledge of marine systems is still so fragmentary that it is difficult even for marine scientists, much less members of the general public, to determine in advance just what dangers a particular human activity poses to marine life. People may have heard brief newspaper reports about chemical contaminants such as mercury in seafood, and they may have seen the remnants of an oil spill on their favorite beach, but their understanding of the full story of chemicals and oil in the ocean is far from accurate or complete.

A case in point is the story of human-introduced mercury in the sea, a story that illustrates the difficulty of both prediction and action in such matters. Mercury, an element that in high concentrations is severely toxic to many organisms, is naturally present in the ocean in reasonably large amounts; marine chemists estimate the world ocean's total mercury content at about 70 million tons. Mercury, like ciguatoxin, is absorbed from sea water by many planktonic organisms, and biological magnification steadily concentrates it in the higher trophic levels. Top-level carnivores such as tuna and swordfish have therefore exhibited mercury levels of nearly 0.4 parts per million (ppm) since the late nineteenth century.

Given the natural mercury content of the oceans, industrial additions of mercury to the sea are small overall and probably of minor global importance. Locally high mercury concentrations are a different matter. Mercury, which is required as a catalyst in a variety of chemical reactions, is released into rivers and estuaries by chemical plants around the world. In 1938, a plant in Minamata Bay, Japan, began discharging mercury into the sea. Because the mercury was released in a form that was water-insoluble, plant officials and local authorities probably gave it little thought, imagining that the compound would simply sink to the bottom of the bay and stay there.

The mercury did sink to the sediment surface as expected, but then it was unexpectedly altered chemically into methyl mercury by bacteria in the sediments. The methyl mercury dissolved easily in the water and at once began rising through the food chain, accumulating at levels of between 10 and 55 ppm in the fish consumed in large quantities by local fishermen. By the mid-1950s, people nearby began to lose their sight, hearing, and coordination. Some were unable to speak. Others regressed into infantile helplessness. Over 100 people were affected, and as many as 50 of them died from what became known as Minamata disease. After a similar incident in Niigata, the Japanese government recognized that mercury poisoning was the cause of the problem and closed the plants.

The stage was set. In 1969, mercury levels between 1.6 and 5.0 ppm were found in Great Lakes pike. Because local residents in both the United States and Canada ate far less fish than the Japanese of Minamata, no human poisonings were reported. Nonetheless, the findings triggered a panic. All fishing in the area was stopped, and the United States Food and Drug Administration announced that existing levels of mercury pollution represented "an intolerable threat to the health and safety of Americans." Safety standards for mercury content were hurriedly set, and quick testing of canned fish samples resulted in removal of more than a million cans of tuna from the market. The tuna industry reeled under the impact.

Within a year, closer examination of available data led to establishment of different, less stringent safety guidelines. Most of the previously banned tuna was put back on the market, although swordfish—which was also included in the original ban—remained prohibited.

The FDA clearly acted on the best available evidence in issuing the original ban; it was forced to decide speedily, on the basis of fairly limited data, what move was best to protect the public interest. In this case, the agency seems to have erred initially on the side of safety, and the decision ultimately damaged both the tuna industry and the FDA's credibility.

But it is still not certain how scientists, physicians, and policymakers can best help such agencies make accurate judgments in the future. The procedure for establishing health guidelines is hotly debated in both political and scientific circles. The topic of permissible contamination levels is an important one because biological magnification significantly concentrates numerous industrial and agricultural chemicals, from DDT to PCBs, even when initial pollutant concentrations are very low. Once toxic chemicals accumulate in higher trophic levels, they persist for long periods even when the original source has been eliminated. Unfortunately, many equally important (if not more important) decisions on food and water contamination must be made on the basis of even less data than were available in the case of mercury.

Petroleum operations and oil spills present different problems. Questions remain about just how harmful normal oil operations and routine, small oil spills really are to marine life. Oil company scientists are quick to point to increases in fish populations around many offshore oil rigs: the submerged metal structures provide a hard substrate for algae and filter feeders, helping to attract fish normally associated with rocky areas that would otherwise not be found far from shore. But other impacts of oil exploration and transportation are not so benign. During the drilling process, finely powdered material called drilling mud, which is used to lubricate the drill, is discharged into the water around the rig. The effect of this extra sediment on both bottom-dwelling and planktonic organisms is not yet known. Onshore terminal development in mangrove areas, near-shore dredging operations, and offshore introduction of drilling mud all change water circulation patterns and increase the water's load of suspended material. Some organisms can tolerate this extra rain of silt and peat, but others, including many corals, are accustomed to clear water and have trouble cleansing themselves.

Then there is the question of oil spills themselves. Open-water spills do not directly affect bottom communities very much, but can cause abnormalities or death among plankton. Because plankton populations in places such as George's Bank are both home and food for commercially valuable fish and shellfish, repeated spills and repeated discharges of drilling mud might have unforeseen results on heavily fished stocks. A great deal more research is necessary before scientists sufficiently understand the effects of oil spills in such situations.

When oil washes into a salt marsh or mangrove swamp, on the other hand, the effects are immediate and obvious. Massive kills of up to 95 percent of the bottom-dwelling organisms usually occur within days. These kills strike public sensibilities the hardest: helpless seabirds dripping with oil, crabs staggering around and collapsing, mussels agape and rotting at low tide. But evidence of this mass mortality lasts only for a few days or weeks. Animals die, decompose, and are carried out by the tide. Their smells and their carcasses disappear. Soon, a television crew could enter the marsh and take a few shots of pristine-looking tidal flats at sunset, and a viewer could hardly believe anything was wrong.

But other, longer-lasting consequences of oil intrusion are not very obvious. Microorganisms in the sediments, vital links in the marsh's nutrient cycles, become coated with hydrocarbons. Some bacteria alter their metabolism, some plants die, and some—the natural hydrocarbon decomposers—proliferate. Sea-grass populations smother at their base, out of sight below the lowest tides, and the sediments they previously held in place begin to shift.

Animals such as lobsters and catfish, which depend on their sensitive chemical senses to inform them of what to eat and when and with whom to mate, suffer from gustatory hallucinations; confused, they prefer oil-soaked debris to their natural foods, and mating success suffers. Eggs and the planktonic larvae of organisms whose adults readily survive the presence of a little oil become fouled, develop abnormally, die, or refuse to settle and metamorphose.

Oil mixes with loose sediment and slowly oozes down tidal channels, into the estuary bottom and beyond. Spreading slowly for months after a spill, oil and oil-soaked sands may remain for as long as several years, preventing the growth of any but the most pollution-tolerant organisms.

Most insidious of all is the low-level, long-term uptake of hydrocarbons by animals and plants. Hydrocarbons accumulate in living tissues, where they may cause slight (or not so slight) changes in metabolic processes. Mussels not obviously harmed by a spill may become sterile. In young animals of all sorts, hydrocarbons synergize with other sources of stress to cause significantly more damage than either stress would cause alone. Some particularly carcinogenic, colorless, odorless, tasteless hydrocarbons persist in animal tissue for months.

Ultimately, fouled areas probably do recover from the effects of a spill, and isolated spills may have little permanent ecological effect. But whether on George's Bank, in the Gulf of Mexico, in the Red Sea, or off the coast of Australia, repeated incidents are bound to cause trouble. Oil spills in coastal waters, it should be noted, already top 100 million barrels each year.

Unfortunately, as ecologist William Murdoch wrote, "large corporations not only have the power to pollute, they have the economic and political power to prevent, delay, and water down regulatory legislation. They also have the power and connections to ensure that the regulatory agencies don't regulate as they ought to." To that impressive list of powers one might add that big corporations have sufficient resources and talented public relations departments to confuse and mislead an underinformed public.

Yet the problems of pollution and habitat destruction are hardly restricted to capitalistic societies. The bureaucracy that controls industrial development in the Soviet Union seems just as deaf to the pleas of biologists as major corporations in the United States are. Lake Baikal, the deepest lake in the world, is in serious trouble from pollution, and the Caspian Sea is, in places, little better than an open sewer for industrial wastes. The approach that China and Third World nations will take toward the natural environment as they hurtle into industrialization has yet to be seen. China's long experience with the ecological results of dense human settlement offers hope that it will proceed with caution. In many Third World countries, however, severe economic crises couple with skyrocketing human populations to impel governments toward drastic, short-term solutions that seriously compromise water pollution controls and undermine the integrity of vital coastal ecosystems.

In Western democratic society, at least, responsibility for pollution rests not only with international corporations and the government, but with every citizen. It is an easy, self-satisfying exercise to point an angry finger at the sneering corporate scoundrels of satirists' cartoons who twiddle their handlebar moustaches as they bulldoze coastal wildlife sanctuaries and pour toxic chemicals into estuaries. But it is the voting public that empowers elected and appointed officials to whom "environment" and "conservation" are dirty words and to whom pollution control represents little more than an obstacle to economic growth. Consumers in free-market societies purchase goods and services provided by pollution-creating industries, and, until recently, consumers have been unwilling to shoulder the cost of pollution control associated with those industries.

There is, unfortunately, no simple fix. Pollution control has its costs, and under the free-market system, it is the public that must pay those costs. No sensible environmentalist today would argue that cleaning up polluting industries and equipping new ones with environmental safeguards is cheap. Even preserving wilderness areas seems expensive; offshore wildlife sanctuaries generate no tax revenues and—aside from preserving industries such as fisheries that most people away from the coast tend to take for granted—generate far fewer jobs than oil and gas exploration and development.

But long-term costs of marine pollution far outweigh the short-term economic benefits to be gained by ignoring proper controls. No one would suggest that halting progress, ceasing oil exploration, or shutting down every polluting industrial operation is a viable option. But research into long-term environmental effects is desperately needed. Caution is essential in directing development to avoid expensive, inconvenient, and occasionally health-threatening environmental mistakes. For every penny saved now by ignoring the need for pollution control, people will pay back many dollars in the future. Critical spawning and nursery areas, from coastal estuaries to offshore banks, were not selected by conservationists. They were adopted by marine organisms in the course of their long and wandering evolutionary relationships with one another and with their environment.

Human beings cannot choose which areas are ecologically important; we can only do our best to identify those we must protect from befoulment. Like the evolutionary interactions among all other organisms in the biosphere, the relationship of *Homo sapiens* with the sea and its inhabitants has been a long and wandering one. We have made mistakes, we have hit several dead ends, and we have made several surprisingly felicitous discoveries. But our power as intelligent and accomplished beings makes the future course of our wanderings critically important. Mindful of the weight of our immense numbers, we must tread lightly as we wander, and we must examine our options at each juncture before rushing ahead. The sea, although far from bottomless, can provide food and drugs in abundance. We can, however, maintain and expand our harvest of the oceans' bounty only if we understand and preserve the beautiful, resilient, yet vulnerable habitats of the diverse marine organisms that constitute the oceans' living legacy to mankind.

Epilogue

Threatened and Endangered Marine Species

Humankind's relationship with the rest of the living world is of increasing concern to all of us who study life on earth. Our ability to disturb permanently the balance of life both on land and in the sea is mounting at a frighteningly rapid rate. Earth is currently in the throes of a biological upheaval unprecedented in the entire history of life on this planet, as human-caused disturbances force animal and plant species into extinction at an ever-increasing rate. Before the turn of the century, nearly 1 million of the approximately 5 to 10 million species estimated to inhabit the earth today will probably have vanished. If humankind continues on the same course, somewhere between one-third and one-half the world's species will have disappeared forever within the next 500 years.

The more prescient members of the scientific community have been warning the world about these matters for some time now, but the situation continues to worsen. Steadily growing human populations demand ever more in the way of space, food, and energy from a finite world. Along with several terrestrial habitats under severe stress from human activity, time is rapidly running out for many plants and animals inhabiting the oceans' vulnerable borders: the overfished, overcrowded, and pollution-threatened continental shelves. As you have seen, it is here that most of the ocean's

Hawaiian monk seal
(courtesy U.S. Fish & Wildlife Service)

Estuarine crocodile
(© Don & Pat Valenti/Tom Stack & Assoc.)

Following pages:
Manatees
(© Douglas Faulkner)

bounty lies. It is also here that man's impact is most powerful.

In an effort to broaden the base of support for intelligent planning and exploitation of marine resources, we have throughout the book repeatedly taken what some may view as a rather mercenary approach—emphasizing the value of marine life in economic terms. For scientists, naturalists, and those who share their world view, the natural beauty of salt marshes, coral reefs, and kelp forests needs no justification; no earthly price tag could ever measure their value. These lucky people include in their definition of their own humanity an intimate philosophical and emotional relationship with the wondrous creatures whose planet we share.

Unfortunately, such people are a minority in modern society. Everyone else wants to know "What's in it for me?" Business and government, both in developed countries and in the Third World, constantly defer to the bottom line. If these people are to be convinced that environmental preservation is necessary, or even desirable, they are far more likely to respond to clear economic arguments than to talk of moral imperatives. So, responding to the rallying call of outspoken scientists like Norman Meyers, our text has reminded you—sometimes gently and sometimes not so gently—just how much the good health of marine life ultimately means to humanity in dollars and cents.

The accompanying pictorial epilogue highlights just a slight few of the marine species whose populations are threatened by human activities or whose numbers are already so small that they are in imminent danger of extinction. Those to whom all the preceding talk of economic incentives seems cynical or hard-hearted need only look at these photographs and the others in the book for reminders that, even if they do have prices upon them, marine organisms will always be beautiful.

Green sea turtles
(© Bob McKeever/Tom Stack & Assoc.)

Brown pelican
(© Steven C. Kaufman/Peter Arnold Inc.)

Selected Readings

To Live in the Sea. One of the more readable texts offering a process-oriented overview of marine ecosystem form and function is *Fundamentals of Aquatic Ecosystems,* by Barnes and Mann (Boston: Blackwell Scientific, 1980). For an illustrated guide to marine organisms and their habitats, try either McConnaughey and Zottoli's *Introduction to Marine Biology* (St. Louis: C. V. Mosby, 1983) or Sumich's *An Introduction to the Biology of Marine Life* (Dubuque, Iowa: Wm. C. Brown, 1984). And for an introduction to the world of marine primary producers, look at Dring's *The Biology of Marine Plants* (London: Edward Arnold, 1982).

Steven Jay Gould explores the world of evolution in a series of intriguing, expertly written personal perspectives on the evolutionary process and evolutionary theory. These books—*Ever Since Darwin* (1977), *The Panda's Thumb* (1980), and *Hen's Teeth and Horse's Toes* (1983)—are all published in New York by Norton. Gould and his colleague Richard Lewontin also launched an attack on certain sloppy aspects of Neo-Darwinian thinking; see their 1979 article "The Spandrels of San Marcos and the Panglossian Paradigm: A Critique of the Adaptationist Programme," *Proceedings of the Royal Society of London Bulletin* 205:581–598.

A number of cognoscenti offer their views on evolutionary process and theories in *Scientific American* magazine's special issue on *Evolution,* September 1978, 239:3.

Where the Sea Meets the Land. Ecologists John and Mildred Teal offer a personal, popular, readable, yet scientifically valid account of salt marsh history, significance, and interactions with humankind in *Life and Death in a Salt Marsh* (Boston: Atlantic, 1969). A collection of topical, accurate, and highly readable articles on marshes and similar ecosystems can be found in *Oceanus* magazine's special issue on *Estuaries,* Fall 1976, 19:5.

Underwater Producers and Consumers. The complex physical and biological phenomenon of upwelling is well covered in *El Niño,* a special issue of *Oceanus* magazine (Summer 1984, 27:2).

Capriccio on a Coral Reef. For a detailed survey of coral reef organisms, their ecology, and a bit of their history, try Endean's *Australia's Great Barrier Reef* (New York: University of Queensland Press, 1982). And German biologist Hans Fricke provides a very readable, illustrated account of coral animals and their behavioral interactions in his *The Coral Sea* (New York: Putnam's, 1973).

Water Colors. Two fairly technical volumes on the generation of color in marine organisms (accessible to those with some background in science) are *Cromatophores and Color Change* by Bagnara and Hadley (Englewood Cliffs, N.J.: Prentice-Hall, 1973) and Fox and Vevers, *The Nature of Animal Colours* (London: Sidgwick and Jackson, 1960). A more popularly written account of one animal's camouflage techniques is provided in A. Packard, "What the Octopus Shows to the World," *Endeavor* XXVIII:104, 92–99. For a discussion of the interactions among aquatic visual environments, color vision in aquatic animals, and color communication, see "Visual Communication in Fishes" by Levine, Lobel, and MacNichol (pages 447–475 in *Environmental Physiology of Fishes,* edited by Ali, New York: Plenum, 1980) and Levine and MacNichol, "Color Vision in Fishes," *Scientific American,* 246(2): 140–149.

For tantalizing glimpses into other senses of marine organisms, see the *Oceanus* special issue on *Senses in the Sea* (Fall 1980).

Poison, Tooth, and Claw. The masterwork in the field of marine toxins is Halstead's physically intimidating yet surprisingly readable *Poisonous and Venomous Marine Animals of the World* (Princeton, N.J.: Darwin Press, 1978).

To Exhaust the Sea. For humbling, occasionally depressing, yet positively oriented assessments of human interactions with the biosphere past, present, and future, try A. J. Fritsch's *Environmental Ethics* (Garden City, N.Y.: Anchor, 1980) and N. Meyer's *A Wealth of Wild Species* (Boulder, Co.: Westview, 1983).

The most comprehensive work in the area of aquaculture is *Aquaculture,* by Bardach, Ryther, and McLarney (New York: Wiley-Interscience, 1972). The complex biological, oceanographic, legal, and economic issues of world fisheries are explored in two special issues of *Oceanus: Harvesting the Sea* (Spring 1979, 22:1) and *The Exclusive Economic Zone* (Winter 1984–85, 27:4).

Index

Page numbers in *italics* indicate illustrations

Acknowledgments

To Richard Peterson for goading me into and guiding me through my first popular articles; to Chester Roys and Norton Nickerson for first leading me down the coral-garden path; to my mentors since then for their gifts of knowledge and perspective; to Maxim Daamen for hospitality at Tiverton-by-the-sea; to Ivan Valiela, David Caron, and Leslie Kaufman for comments on chapters 2, 3, and 4; to my colleagues for their enthusiasm and willingness to share the fruits of their labors; to my students at Boston College for joy and inspiration; and to Sheila, Dick, Debbie, Rich, Greg, Billy, Mark, Steve, and Richard for the affection, support, and diversions of friendship.

To all these people, sincerest thanks.

I am rarely alone in the sea. I have shared my time with some great dive buddies whose company has brought many special moments over the years, and whom I want to thank: Ken Beck, who talked me into my first wet suit; Amos Goren, a master of south Sinai's reefs; Yair Harel, who shared with me his curiosity and enthusiasm for diving; Muhamed Hagrass, who explored "The Brothers" with me; Roy Hauser and the "Truth" for showing me California's kelp forests; Bob Johnson, for his company in my deepest dives; Jon Kenfield, for his British precision; Brian Lyons, whose exploits on New England's wrecks taught me the nuts and bolts of diving; Ernst Meier, for sharing my escapades; Urs Mockli, the Swiss marine mammal, who led me to some great adventures; Mark Schoene, who inched his way beside a torpedo ray for my camera; Laurie Seluk, for wrestling with a giant lobster; Galit Rotman, my favorite dive and life partner.

As this book has taken shape over the years, two very special people have made substantial contributions to which no acknowledgment could possibly do justice. Our heartfelt thanks to both: to Barney Karpfinger for help and encouragement far exceeding expectations of both agent and friend; and to Roy Finamore for patience, wit, insight, and editorial acumen.

Joe Levine and Jeff Rotman

Undersea Life

Designer: J. C. Suarès
Assistant Designer: Gordon Harris

Composed in Times Roman and Futura Medium
by Arkotype Inc., New York, New York
Printed and bound by Weber Colour Printing, Ltd.,
Bienne, Switzerland